SPECTROSCOPY AND ITS INSTRUMENTATION

P. BOUSQUET
Professeur à la faculté des Sciences de Marseille

Spectroscopy and its instrumentation

translated by
K. M. GREENLAND
Sira Institute

Preface by
P. JACQUINOT
Membre de l'Academié des Sciences

Adam Hilger
London

This volume is a translation of *Spectroscopie Instrumentale*, published by Dunod, Paris

© K. M. Greenland, 1971
ISBN 0 85274 180 4

Published by
ADAM HILGER
Rank Precision Industries
31 Camden Road, London, NW1 9LP

Made and printed in Great Britain by
William Clowes & Sons, Limited
London, Beccles and Colchester

PREFACE

by

P. JACQUINOT
Membre de l'Académie des Sciences
Professeur à la Faculté des Sciences d'Orsay
Directeur-Général du C.N.R.S.

A prism or grating with a few optical elements and an appropriate receiver (the observer's eye, a photographic plate or a photodetector) make up a deceptively simple instrument for observing or recording optical spectra, used over the years by countless numbers of physicists, chemists and biologists. Yet few will have given thought to the best way of operating these instruments, or to the practical limit of performance of which these simple assemblies of simple components are capable. It is nevertheless true that a treatment of these problems based on fundamental and mutually independent optical parameters leads to a knowledge of spectroscopic properties and possibilities that not only is of great practical value but sometimes even contradicts accepted practice based on ill-founded premises. This method of approach also facilitates an understanding of the characteristics of more complex systems such as the interferometric instruments, brings out their underlying relationship with the 'classical' methods, and leads to the invention of new instruments of outstanding performance.

A study of the principles of spectroscopic instrumentation on these lines has been incorporated by some universities in their special advanced courses—an invaluable feature of the French educational system known as *le troisième cycle*—notably in Paris for some time past and, more recently, in Marseille.

Dr P. Bousquet, a Professor of the Faculty of Science of Marseille, who has lectured on the principles of spectroscopy in the *troisième cycle* for several years, was eminently qualified to write a book on the subject with which he has been so intimately associated in this course. His book is timely and answers a real need; I know of no other work, either of the present day or, indeed, of any period in the evolution of optical spectroscopy, which has so firmly laid emphasis on the principles and scope of the basic methods of spectroscopy, thus avoiding the danger of concentrating on ephemeral details of instrument design. The author has not only steered clear of this hazard but has also avoided the temptation to be drawn into abstruse theoretical discussions which would have been inappropriate in the context of this book.

Professor Bousquet has succeeded in striking the right balance between the presentation of well-established methods and those that are 'new'. His book is clearly written and straightforward, getting down to fundamentals where depth of treatment is called for, and adequately covering a subject which, after a long quiescent period, has recently gone through a phase of rapid evolution. There must now follow a period of dissemination and assimilation of these new principles in which, I am sure, Professor Bousquet's book will play an effective part.

AUTHOR'S FOREWORD

During the last twenty years remarkable advances have been made in the instrumentation of spectroscopy. While conventional techniques are still widely employed and have indeed been constantly improved, entirely new methods have made their entry, greatly increasing the scope and accuracy of measurements.

Among the older methods, grating spectrographs and spectrometers have made outstanding progress through recent advances in methods of producing diffraction gratings, advances due in great part to the work of G. R. Harrison and G. W. Stroke.

Instruments using slits are, however, inevitably limited in performance by their very nature: it is only by the development of fundamentally different methods that this limitation has been overcome. Almost all these new methods are applications of interference spectroscopy and are largely due to the school of P. Jacquinot at the Aimé Cotton laboratory of the Centre National de la Recherche Scientifique. A colloquium held in 1966 at Orsay focused attention on these advances and emphasized the progress made in recent years.

It is not mere chance that so much progress has been concentrated in so short a period of time, with the evolution of novel methods breaking completely with traditional techniques in both principle and application. On the contrary, this is the result of a logical process of analysis of the problem and clarification of the laws governing the functioning of dispersing instruments. Among the concepts brought into prominence by the work of P. Jacquinot, that of *luminosity* is the most important. Although the significance of this quantity has often been overlooked in the past, it is fully comparable with that of the more familiar resolving power: there is no point in being able to isolate very narrow wavebands if their content of energy is too small to be measured. We find, indeed, that in some instruments the signal-to-noise ratio of their output has become an important factor in performance.

These considerations lead to the idea of combining in one function a number of characteristics that together determine the performance of a dispersing instrument; this function, known as the *instrumental profile*, facilitates a true comparison of instruments of differing types.

It was in the context of these new aspects of instrumental spectroscopy that this monograph was planned. Its aim is to present the essential features of all methods in present-day use, with special attention to the more recent of these and to the principles on which each is based. This should lead to an understanding of the reasons for developing each instrument, and so to an appreciation of its utility in its particular area of application.

While this book has its origin in a course of advanced optics for students in the Faculty of Science of the University of Aix-Marseille it has been written in such a way that no specialized knowledge of the optics of spectroscopy is necessary. For this reason some space has been devoted to a brief description of the most important of the classical instruments such as the Fabry–Perot interferometer.

My work enjoys the distinction of a preface by Professor P. Jacquinot, Director of the Centre National de la Recherche Scientifique, who, in spite of his many duties, also found the time to read through my text and to make most valuable comments on it; I wish to express my deep gratitude for his kindness. I also acknowledge my great indebtedness to Professor P. Rouard, Dean of the Faculty of Science in Marseille, who invited me to write this book. Without his unstinted encouragement and advice it would never have seen the light of day: I am happy to have this opportunity to thank him in all sincerity.

TRANSLATOR'S NOTE

Recognizing the value of Professor Bousquet's book to English-reading technologists, scientists and students, I very willingly accepted Professor Bousquet's invitation to undertake its translation. My only regret is that I have been quite unable to match the elegance of his style of writing. I am more confident about the accuracy of the text itself because Dr W. T. Welford, Reader in Applied Optics at Imperial College, London, very kindly read and criticized the translation; I am greatly indebted to him for his help.

Since some of the original key references are to French sources, I enlisted the expertise of Miss Anne Bugden to complement these with their equivalents in English wherever possible: I am grateful to her for undertaking this task. Finally, I must acknowledge the ready co-operation that I have enjoyed with Professor Bousquet himself; it has been a pleasure and a privilege to give his book the wider readership that it well deserves.

LONDON K. M. GREENLAND
May 1971

CONTENTS

1 Introduction — 1

2 General properties of prism and grating spectrographs — 7

3 Prism spectrographs — 28

4 Diffraction gratings — 40

5 Grating spectrographs — 79

6 Prism and grating spectrometers and monochromators — 95

7 Interference spectroscopy — 144

Bibliography — 225

Index — 235

CHAPTER ONE

INTRODUCTION

1.1. *The purpose of dispersing instruments*

The spectrum of the light emitted by a source may be characterized by a function $L(\nu)$ expressing the radiance L of the source as a function of the frequency of the radiation emitted; in the same way an absorption spectrum is characterized by a function $A(\nu)$ relating the absorption factor A of the material transmitting the radiation to the frequency content of the radiation. The purpose of dispersing instruments is to provide the maximum amount of information about the function $L(\nu)$ or $A(\nu)$. From now on we shall treat our subject in terms of emission spectroscopy but the conclusions will as a rule be equally valid for the case of absorption spectroscopy.

1.2. *Definition of spectral quantities*

Wavelength. Wavenumber

It is not possible to make a direct measurement of the frequency of an electromagnetic vibration, but absolute measurements of wavelength present no great difficulty. It is partly for this reason that in spectroscopy it is common practice to describe monochromatic radiations by their wavelength λ rather than by their frequency ν.

The usual unit of wavelength in the optical region is the ångstrom (Å), a unit linked with the M.K.S. system (1 ångström = 10^{-10} metre); in the infra-red region of the spectrum the micrometre (μm) is preferred because of its more convenient magnitude (1 μm = 10^4 Å).

Wavelength λ_v *in vacuo* is related to frequency ν by the relation $\lambda_v = c/\nu$ where c is the speed of light *in vacuo*. In a medium of refractive index n, and in particular in air, the wavelength becomes $\lambda = c/nv = \lambda_v/n$. Since measurements are almost always made in air its refractive index must be taken into account. The difference

$$\lambda_v - \lambda = (n-1)\lambda$$

is significant (1–2 Å in the visible spectrum) and, of course, depends on the temperature, pressure and composition of the atmosphere during the experiment. For this reason it becomes necessary to define a 'normal atmosphere' to which the results of wavelength measurements are reduced. The accepted normal atmosphere is dry air at a pressure of 760 Torr and a temperature of 15 °C, the proportion of carbon dioxide being 3×10^{-4} parts by volume. The corresponding wavelengths are known as *international wavelengths*; it is these that are usually quoted in wavelength tables.

Up to a short time ago the primary wavelength standard was the international wavelength of the red line of cadmium ($\lambda = 6438\cdot4696$ Å) which had on a number of occasions been measured against the standard metre, in particular by Michelson and Benoît in 1892, then by Benoît, Fabry and Perot in 1913. On 14 October 1960, by unanimous agreement of the thirty-two governments represented, the 11th International Conference on Weights and Measures reversed the situation by adopting a new definition of the metre based on the wavelength of a luminous radiation instead of on the length of a material standard. From that date the metre has been defined as: 'the length equal to 1 650 763·73 wavelengths *in vacuo* of the radiation corresponding to the transition between the $2p_{10}$ and $5d_5$ levels of the krypton-86 atom'.

The red line of cadmium has thus yielded its role of primary wavelength standard to the orange line of krypton-86, which now constitutes not only the wavelength standard but also the primary standard of length.

Although it is convenient in experimental work to characterize a monochromatic radiation by its wavelength, in theoretical work frequency nevertheless emerges as the fundamental parameter. Indeed, in the quantum theory of the emission of luminous radiation by atoms it is the frequency v which is given by Planck's formula $E_2 - E_1 = hv$, in which E_2 and E_1 repre-

COMBUSTION INSTITUTE EUROPEAN SYMPOSIUM 1973

edited by
F. J. Weinberg

Academic Press
London and New York

COMBUSTION INSTITUTE EUROPEAN SYMPOSIUM 1973
edited by F. J. Weinberg
Department of Chemical Engineering and Chemical Technology, Imperial College London, England

August 1973, xviii+740 pp., £12.00
0.12.742350.8

This work contains the collected papers of The Combustion Institute European Symposium 1973, held at Sheffield in September of this year.
The Symposium was attended by Professor Glenn C. Williams, the President of the Combustion Institute, and contributors came from non-European as well as European countries. All the papers are concise accounts of original research and the exceptionally short publication delay will make the book particularly valuable in research laboratories, whilst its large information content destines it to become a standard reference work.
Combustion Institute European Symposium 1973 will be of great value to all those working in the fields of combustion, ignition and flame research in industry, research institutes and university departments.

Contents
Elementary and complex combustion reactions. Fire research, including the burning of plastics. Ignition. Ionisation and electrical aspects of flames. Particulates and droplets: Formation, combustion and radiation. Combustion-generated pollution. Turbulent flames and high intensity combustion. Furnace flames. Flame studies.

Academic Press
London and New York
24-28 Oval Road, London NW1, England
111 Fifth Avenue, New York, NY 10003, USA

Registered office:
Academic Press Inc. (London) Ltd.
24-28 Oval Road, London, NW1 7DX
Registered number: 598514 England

sent the energy of the emitting atom in the initial and final states, and h is Planck's constant. Now, frequency is related to wavelength *in vacuo* by the relationship $v = c/\lambda_v$, in which c, the speed of light, is known to an accuracy well below that achieved in the measurements of wavelength (about 1 part in 3×10^6 for the former compared with 1 part in 3×10^9 for the latter). For this reason spectroscopists have long since adopted *wavenumber in vacuo*, ($\sigma = 1/\lambda_v = v/c$) in place of frequency, the wavenumber being proportional to frequency and known to the same accuracy as wavelength. The unit in common use as a measure of wavenumber is the inverse centimetre (cm^{-1}), sometimes termed kayser (K); the millikayser (mK) is also used. The visible spectrum extends roughly from 4000 Å (25 000 K) to 7500 Å (13 300 K). In the green region of the spectrum a wavenumber interval $\Delta\sigma$ of 1 kayser corresponds to a wavelength interval of about 0·25 Å.

Radiance of source

If a particular source emits a discrete series of radiations which, within experimental limits, can be regarded as monochromatic (a line spectrum), the radiance L of the source for each monochromatic radiation is uniquely determined. A radiation can in general only be regarded as monochromatic in so far as the dispersing instrument employed has a limited resolving power; with an instrument of sufficiently high resolving power the same radiation will be seen to be a narrow band of continuous spectrum. By its very nature, the mechanism of emission cannot generate a strictly monochromatic radiation; even in the case of coherent sources (lasers) there is a measurable frequency spread.

If, now, we consider a spectrum containing a continuous distribution of radiance in a given spectral interval, we need a new quantity, the *radiance per unit wavenumber* $L_\sigma = \partial L/\partial \sigma$. The source radiance in an infinitely small wavenumber interval $d\sigma$ is clearly proportional to $d\sigma$, so we have

$$dL = (\partial L/\partial \sigma)d\sigma = L_\sigma d\sigma$$

In this case the spectrum is characterized by the function $L_\sigma(\sigma)$; it is this function that the instrument has to determine.

1.3. *The two basic features of a dispersing instrument*

Resolving power

For the most accurate determination of the function $L_\sigma(\sigma)$ a dispersing instrument should divide the spectral region under survey into small intervals, or spectral elements, of a necessarily finite width $\Delta\sigma$ and should generate for each interval a signal proportional to the radiance ΔL of the source. The appropriate width of the interval $\Delta\sigma$ depends on the complexity of the spectrum and on the nature of the information required about it. The smaller the value of $\Delta\sigma$ the more highly is the spectrum resolved; $\Delta\sigma$ is thus the *resolution* of the instrument or the *resolvable spectral element*. A more commonly used quantity is the *resolving power* (or *resolvance*) defined as:

$$\mathscr{R} = \sigma/\Delta\sigma = \lambda/\Delta\lambda$$

Luminosity†

The performance of a dispersing instrument is not simply dependent on the value of its resolving power. It is not sufficient that the spectrum be divided into elemental wavebands that are as narrow as possible; each of those elements must contain sufficient radiant energy to permit a sufficiently accurate measure of the radiance of the source in that waveband.

The detector receives a quantity of energy proportional to $\Delta L = L_\sigma \Delta\sigma$, ΔL being the source radiance in the resolved spectral interval $\Delta\sigma$. In the case of instruments having a *total flux* detector the coefficient of proportionality (ignoring transmission losses) is the étendue‡ U of the beam accepted by the instru-

---

† *Translator's note.* Luminosité—a term enjoying common usage in spectroscopy with a meaning quite different from that applying in photometry.

‡ *Translator's note.* Étendue: a geometrical invariant of the optical system, in general measured by the product of the area of the field stop and the solid angle subtended by the aperture stop at the field stop. See Fig. 6.4 (p. 104). Hence for a spectrometer,

U = (effective area of collimator) × (solid angle subtended by entrance slit at collimator)
  = (effective area of focusing lens) × (solid angle subtended by exit slit at lens).

Some authors use the term *luminosity* for this factor but in this book *luminosity* is defined in such a way as to take into account the effect of diffraction.

ment (sometimes termed its 'optical acceptance'). The accuracy of measurements depends on the signal/noise ratio at the output of the detector; putting aside, for the time being, the question of the fundamental difference between photon noise and detector noise, it is easy to appreciate the importance of the factor U which, in this case, determines the *luminosity* of the instrument.

The determining factors of luminosity will be different in the case of instruments employing *flux density* detectors (that is, image receivers); furthermore, the relationship will be more complex for certain indirect spectrometric methods. The concept of luminosity nevertheless retains its fundamental importance in all forms of spectroscopy.

1.4. *Classification of types of dispersive instrument; terminology*

Many different systems are used in spectroscopy but the terminology used to describe them has never been properly standardized; we shall adopt here the terms that seem to be in most common use at the present time.

Dispersing instruments are often classified according to the type of dispersive element employed or the physical effect producing the dispersion. Thus we have *prism, grating* and *interference* instruments. As with all systems of classification, this one is essentially arbitrary since underlying the principles on which these instruments operate is, in every case, the phenomenon of interference. This is obvious for the grating, which is simply a multiple-beam interferometer, but it is true also for the prism. In this latter case there is in fact zero-order constructive interference between the emergent rays in a direction which is different for each wavelength, so giving rise to the effect of dispersion. The proposed classification is nevertheless reasonable and convenient since it is based on practical considerations; accordingly interference spectroscopy covers the use of interferometers proper such as the Michelson or the Fabry–Perot, instruments operating usually in the higher orders of interference and technically very different from grating and (still more so) prism instruments. Another classification is derived from the type of detector employed. The detector is an important factor because its nature can significantly modify the properties of an instrument. Unfortunately the terminology of this area of technology can lead to confusion.

Leaving aside the eye itself, there are two classes of detector, namely the flux density detectors, or image receivers, and the total flux detectors. The image receivers record the flux incident at each point on their sensitive surface; they are represented by the photographic plate and the various image converters (particularly A. Lallemand's electronic camera). The total flux detectors are the photoelectric cells (for example, photomultipliers, photoconductive cells), thermal receivers (such as thermocouples and pneumatic detectors) and, generally, detectors that give a signal proportional to the total incident luminous flux.

A terminology often used in practice, though one subject to much controversy, allocates the term 'spectrograph' to those instruments equipped with an image receiver and 'spectrometers' to those using a total flux detector. This system of designation originates in the fact that flux detectors make possible a direct and accurate measurement of the intensity of spectral lines while image receivers are not so well suited to this purpose. A distinction must, however, be drawn between this and an older usage of the word spectrometer by which is understood an instrument capable of measuring the wavelengths, rather than the intensities, of the spectral lines.

Until a much-needed codification of this terminology has been established we shall reserve the name *spectrometer* for the instrument provided with a *flux detector* and *spectrograph* for those having an *image receiver*. As for the term 'spectroscopy', this will be used, very broadly, to indicate the whole range of techniques for the analysis of optical radiations, so justifying the title of the book.

CHAPTER TWO

GENERAL PROPERTIES OF PRISM AND GRATING SPECTROGRAPHS

In accordance with the conventions stated at the end of Chapter 1, this chapter only deals with instruments provided with an image receiver (usually a photographic plate) and having a prism or grating as the dispersing element. The consideration of interference systems is postponed to Chapter 7.

2.1. Basic layout

The optical layout of the majority of prism or grating spectrographs is shown schematically in Fig. 2.1.

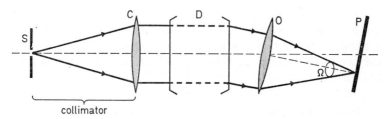

FIG. 2.1. Basic layout of a prism or plane grating spectrograph.

The apparatus consists essentially of:
  A *collimator* composed of the slit S and the collimating lens C
  A *dispersing element* D (prism or plane grating) illuminated by the beam from the collimator
  A *focusing lens* O which forms monochromatic images of

the slit S on a photographic plate P placed in its focal plane.

An instrument of this kind, in which the spectral lines formed on the photographic plate are images of the slit, is free from astigmatism.

There are other types of spectrograph, with layouts differing from that described above, in which the spectral lines are formed by the overlapping of astigmatic focal lines which are the images of elementary lengths of the slit. This happens with spectrographs using concave gratings and with certain rarely used types of prism spectrograph such as the Féry. The astigmatism of these instruments represents a rather serious handicap and they are less commonly used than the stigmatic spectrographs. In this chapter we shall be dealing only with spectrographs of the stigmatic class.

THEORETICAL (OR INTRINSIC) RESOLVING POWER

2.2. *Distribution of radiation intensity in the plane P when the slit is infinitely narrow and the incident radiation is perfectly monochromatic*

The resolving power that may be expected with a spectrograph equipped with a particular dispersing element (prism or grating) is fundamentally limited to a maximum value $\mathscr{R}_0$ by the effects of diffraction. When the slit is infinitely narrow and the aberrations negligible, the distribution of radiation intensity in the plane P is in fact determined solely by the laws of far-field diffraction.

Assuming that the lenses C and O do not affect the beam width (a necessary condition if the performance of the dispersing element D is to be fully exploited), the dispersing element itself acts as the limiting diaphragm. The projection of its outline on a plane normal to the emergent beam is most commonly a rectangle: in Fig. 2.2, a is the width of this rectangle measured in the direction of dispersion—that is, the direction perpendicular to the slit.

GENERAL PROPERTIES OF SPECTROGRAPHS 9

Under the above conditions (infinitely narrow slit, illuminated by a monochromatic radiation of wavelength λ, negligible aberrations, rectangular pupil) the radiation intensity in the

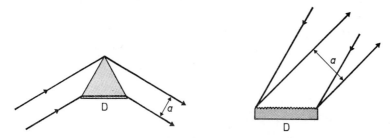

FIG. 2.2. Limitation of emergent beam.

plane P is constant along any line parallel to S† and varies along an axis Ox perpendicular to S as indicated by the curve of Fig. 2.3.

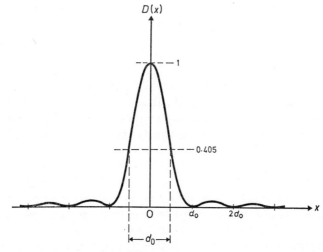

FIG. 2.3. Variation of intensity in plane P when the slit is infinitely narrow and the incident radiation is monochromatic. Axis Ox is perpendicular to the slit; $d_0 = f\lambda/a$.

† Strictly speaking, this assumes an infinitely long slit but is also for practical purposes true for a slit of finite length except in the immediate neighbourhood of its ends.

This curve is composed of a very narrow central region flanked by secondary maxima decreasing very rapidly in intensity. The width of the central region of the diffraction pattern is conventionally measured by the quantity $d_0 = f\lambda/a$ (f being the focal length of the objective O), that is, by the difference of the abscissae of the central maximum and of the first minimum. It should be noted that d_0 is not the width at half the maximum intensity but at the ordinate 0·405.

2.3. *Criterion of resolution*

On this definition of the width d_0 depends the criterion for calculating the theoretical resolving power of a dispersing element and, more generally, the limit of resolution of any optical instrument having a rectangular aperture and using incoherent radiation.

Suppose the slit S, infinitely narrow as before, to be illuminated simultaneously by two monochromatic radiations λ and $\lambda+\Delta\lambda$. In the plane P each of these produces a distribution of flux identical with that of Fig. 2.3; since the two beams are incoherent, the total intensity is the sum of that due to each radiation. It is generally accepted that separation can be detected if the centres of the two patterns are spaced apart by a distance not less than d_0. At the limit of detection, therefore, the central maximum of one is superimposed on the first minimum of the other. The resulting intensity then varies as indicated in Fig. 2.4.

This resolution criterion is of course quite arbitrary, but it cannot be otherwise. The ability to separate two spectral lines or two close images depends on so many parameters that it is not possible to make an absolute rule. In the present case it is in particular a function of the relative intensities of the two radiations; it has been assumed that these intensities are not very different, but it is obvious that if this is not so the criterion given above is no longer valid, since the weaker line may be masked by the spread of the diffraction pattern due to the stronger line. Nevertheless the conventional criterion generally gives values for the limit of resolution agreeing sufficiently well with experiment for it to be very widely utilized, especially as it is so easily applied.

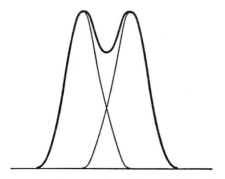

FIG. 2.4. Distribution of intensity resulting from superposition of the diffraction patterns due to two radiations of neighbouring wavelengths when the centres of these patterns are separated by d_0 (Rayleigh criterion).

2.4. *Expression for the intrinsic resolving power of a dispersing element*

For two radiations differing in wavelength by $\Delta\lambda$ there are two corresponding diffraction patterns of which the centres are separated on the photographic plate by

$$\Delta x = f \frac{d\theta}{d\lambda} \Delta\lambda$$

$d\theta/d\lambda$ representing the angular dispersion due to the element. These two spectral lines therefore appear to be separate if

$$f \frac{d\theta}{d\lambda} \Delta\lambda \geq f \frac{\lambda}{a}$$

The spectral interval resolved is therefore

$$\Delta\lambda = \frac{\lambda}{a \, d\theta/d\lambda}$$

and the theoretical (or intrinsic) resolving power of the dispersing element is:

$$\mathcal{R}_0 = \frac{\lambda}{\Delta\lambda} = a\frac{d\theta}{d\lambda} \qquad (2.1)$$

INFLUENCE OF THE WIDTH OF THE SLIT ON RESOLVING POWER AND LUMINOSITY

2.5. *Importance of nature of slit illumination*

Assuming for the time being that aberrations are negligible, the next problem is to calculate the distribution of intensity of radiation in the image of the slit S when the latter has a finite width l. The slit can be regarded as a combination of an infinite number of infinitely narrow slit elements, each giving rise to a corresponding diffraction pattern of width d_0 in the image plane P. To calculate the intensity produced at a point in the plane P by the whole slit S it is necessary to know how to combine the radiations from any two points in S which, by reason of the overlapping of their two diffraction patterns, reach that point. It is the method by which the slit is illuminated that determines the process of combination. In this context, two extreme cases may be defined, between which fall the generality of illuminating systems used in practice.

(*a*) If there is no relation between the radiations emitted by two points of an object, the illumination is said to be perfectly *incoherent*. Under these conditions the *intensities* of illumination must be summed at each point in the image plane.

(*b*) If, on the contrary, the phase difference and the relationship of the amplitudes of the flux emitted by two points of the object remain constant the condition is one of perfectly *coherent* illumination. Here the *complex amplitudes*, not the intensities, must be summed at each point in the image plane. The resulting distribution of intensity will then be found to be significantly different from that observed in incoherent illumination.

(*c*) The practical condition of illumination always lies

GENERAL PROPERTIES OF SPECTROGRAPHS

between these two extremes and is one of *partial coherence*. In spectroscopy the most usual arrangement is to form an image of the source on the slit. The degree of partial coherence in this image depends only on the angular aperture of the beam incident on the slit. Coherent illumination would correspond to an almost zero aperture while complete incoherence would require an infinitely large aperture, which is obviously impracticable; nevertheless, the degree of coherence at the slit is generally so low that no significant error is introduced if it is treated as being completely incoherent.

2.6. Calculation of intensity $E(x)$ at a point in the image of the slit

We take rectangular axes $\omega\xi$, $\omega\eta$ in the plane of the slit oriented so that $\omega\xi$ is perpendicular to the slit, and axes $0x$, $0y$ in the image plane P optically conjugate with $\omega\xi$ and $\omega\eta$ (Fig. 2.5). The image of ω coincides with 0 and the units of length along $\omega\xi$ and $0x$ are chosen so that the abscissae ξ and x of two conjugate points have the same numerical value. (In other words we take the magnification in the direction $0x$ as unity.) †The incident radiation is assumed to be monochromatic.

Let us take first of all an infinitely narrow slit coinciding with the axis $\omega\eta$. This gives rise to a diffraction pattern in the plane P of width $d_0 = f\lambda/a$ having its central maximum on the axis $0y$. Along the axis $0x$ the intensity is distributed as shown in Fig. 2.3, which is derived from the expression

$$D(x) = \frac{\sin^2(\pi x/d_0)}{(\pi x/d_0)^2} \qquad (2.2)$$

Correspondingly a slit element of width $d\xi$ at the abscissa ξ in the object plane generates a diffraction pattern identical with that at O but for which the central maximum has an abscissa $x = \xi$ (Fig. 2.5). The intensity at any point x in this diffraction

---

† This magnification may in practice differ from unity for two reasons:
(a) Different focal length of the lenses C and O.
(b) Existence of *anamorphism* (magnification differing in the $0x$ and $0y$ directions). This happens with a prism when it is not in the position of minimum deviation or with a grating when the angles of incidence and diffraction are different.

pattern is clearly proportional to $d\xi$ and, from (2.2), may be written

$$dE = k\,D(x-\xi)\,d\xi = k\,\frac{\sin^2\left\{\dfrac{\pi(x-\xi)}{d_0}\right\}}{\left[\dfrac{\pi(x-\xi)}{d_0}\right]^2}\,d\xi$$

The constant k depends on, among other parameters, the radiance L (assumed to be constant) of the slit, on the aperture of the optical system, and on the units of measurement.

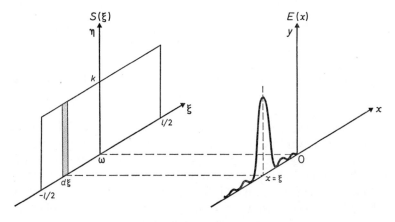

FIG. 2.5. Formation of the image of a slit of width l.

Since the illumination of the slit is assumed to be incoherent, the intensity $E(x)$ due to the whole width l of the slit is obtained by summing the intensity due to each elementary slit:

$$E(x) = k\int_{-l/2}^{l/2} D(x-\xi)\,d\xi$$

$$= k\int_{-l/2}^{l/2} \frac{\sin^2\left\{\dfrac{\pi(x-\xi)}{d_0}\right\}}{\pi\left(\dfrac{x-\xi}{d_0}\right)^2}\,d\xi \qquad (2.3)$$

$E(x)$ may also be written:

$$E(x) = \int_{-\infty}^{+\infty} S(\xi)D(x-\xi)\,d\xi$$

GENERAL PROPERTIES OF SPECTROGRAPHS

in which $S(\xi)$ is a function having the value k in the region $(-l/2, l/2)$ and zero outside this region (Fig. 2.5).† This function $S(\xi)$, the 'slit function', describes (when given an appropriate coefficient) the distribution of flux in the object plane; $E(x)$ therefore represents the general form of the expression giving the distribution of intensity in the image of an incoherently illuminated extended object. The expression shows that the intensity at a point in the image plane is equal to the convolution of the function $S(\xi)$ representing the distribution of intensity in the object and the function $D(x)$, the distribution of intensity in the image of a slit element, or, in symbols, $E = S * D$.‡

After a change of variable $(x - \xi) = X$ the expression for the intensity distribution becomes

$$E(x) = k \int_{-l/2}^{l/2} D(x-\xi)\, d\xi = k \int_{x-l/2}^{x+l/2} D(X)\, dX \qquad (2.4)$$

Integration by parts then gives

$$E(x) = k \int_{x-l/2}^{x+l/2} \frac{\sin^2(\pi X/d_0)}{(\pi X/d_0)^2}\, dX$$

$$= \frac{kd_0}{\pi} \left\{ \left[\frac{\sin^2(\pi X/d_0)}{\pi X/d_0} \right]_{x+l/2}^{x-l/2} + \int_{x-l/2}^{x+l/2} \frac{\sin(2\pi X/d_0)}{2\pi X/d_0} \cdot \frac{2\pi}{d_0}\, dX \right\}$$

The final integral may be calculated with the aid of a table of values of sine integrals so that it is possible to draw curves representing $E(x)$ for various values of the width of the slit. But a simple graphical method of evaluating the function $E(x)$ is also possible, since it is only necessary to measure the area enclosed between the curve $kD(X)$, the X-axis and the two

---

† In terms of a currently adopted notation, this function may be written $S(\xi) = k\,\mathrm{rect}(\xi/l)$. $y = \mathrm{rect}\, x$ (rectangle x) signifies a function defined as follows
$$y = 1 \text{ if } |x| < \tfrac{1}{2}, \quad y = 0 \text{ if } |x| > \tfrac{1}{2}$$

‡ In the general case of any optical instrument, the radiance in the object plane varies in two dimensions and the function D (or *spread function*) represents the distribution of intensity in the image of a single point.

perpendiculars $X = x \pm l/2$ (Fig. 2.6). In this way the change of $E(x)$ with the width l of the slit is readily seen.

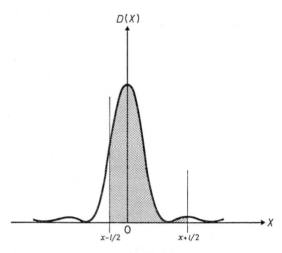

Fig. 2.6. Graphical determination of the convolution $E(x) = S(\xi) * D(x)$.

2.7. *Variation of distribution of intensity in the image with slit width*

The curves of Fig. 2.7 show the trend of the image characteristic as the slit widens. We see that the ordinate E_M of the maximum of $E(x)$, that is, the intensity at the centre of the image, increases almost in proportion to the width l,† provided that the ratio l/d_0 stays less than unity. Above that value the increase is very slow and the intensity at the centre tends to a limit E_0. The reverse is the case for the width d of the image,‡

† It must be remembered that the magnification in the direction $0x$ is always taken to be unity. If the magnification is not unity l may be regarded as representing, not the slit width, but the product of this width by the magnification factor, that is, the width of its geometrical image (the width that it would have if diffraction did not exist).

‡ d is not the width at half the peak value of the characteristic curve for $E(x)$ but the width measured at the ordinate $0.405 E_M$; the purpose of this convention is to facilitate comparisons with the results obtained when the slit is infinitely narrow.

GENERAL PROPERTIES OF SPECTROGRAPHS

which remains practically constant and equal to d_0 while l/d_0 is less than unity but then grows rapidly as l increases.

The general form of these results is readily explained as follows. If l is small compared with d_0, the expression (2.4) for $E(x)$ simplifies to

$$E(x) \approx kl\, D(x)$$

The characteristic curve of the function $E(x)$ is in this case

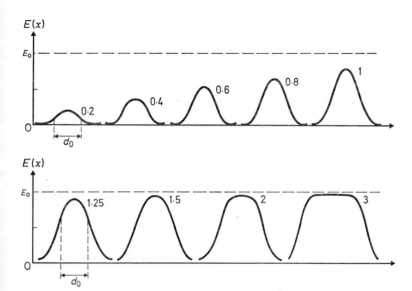

FIG. 2.7. Dependence of distribution of intensity in the image on slit width. The value of the ratio l/d_0 of slit width to width of diffraction pattern due to an infinitely fine slit is shown beside each curve.

derivable from the curve representing $D(x)$. Its width remains the same as for $D(x)$ but the ordinates of $D(x)$ are multiplied by kl.

When the width l of the slit becomes large compared with d_0, it is to be expected that diffraction would only have a negligible influence and that, in consequence, the form of the image would be determined by the rules of geometrical optics. Under these conditions, the width of the image is, nearly enough, equal to l and the distribution of intensity is uniform, independent of l, and calculable by means of the laws of

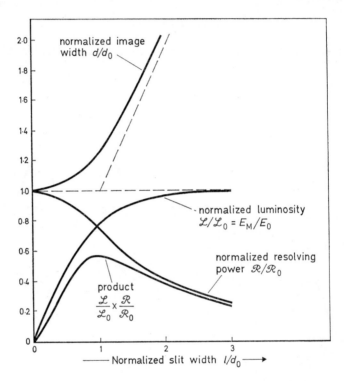

FIG. 2.8. Influence of slit width on resolving power and luminosity of a spectrograph (after P. JACQUINOT and CH. DUFOUR, *J. Rech. Cent. nat. Rech. Scient.* 1948–9, **2**, 91–103).

photometry. Let E_0 be this limiting value of intensity; from (2.4) we have:

$$E_0 = k \int_{-\infty}^{+\infty} \frac{\sin^2(\pi X/d_0)}{(\pi X/d_0)^2} dX = k \frac{d_0}{\pi} \int_{-\infty}^{+\infty} \frac{\sin^2 u}{u^2} du = kd_0$$

(putting $\pi X/d_0 = u$).

Expressing E_0 as a function of the radiance L of the slit, of the transmission factor τ of the instrument and of the solid angle Ω of the emergent beam (Fig. 2.1) we may write as a good approximation:

$$E_0 = L\tau\Omega \qquad (2.5)$$

The preceding results may be summarized by a graphical presentation (Fig. 2.8) of the variation of (i) the intensity E_M at the centre of the image and (ii) the width d of the image, as a function of the width of the slit. It is convenient to use normalized coordinates; the abscissae are in units of l/d_0 and for the ordinates the width d of the image is expressed as a fraction of d_0.

2.8. Influence of slit width on resolving power and luminosity

Definition of luminosity

The photometric quantity determining the blackening produced at a point on the photographic plate is the intensity E at that point. It is therefore logical to define the luminosity of a spectrograph by the ratio of the intensity at the centre of a spectral line to the radiance L of the slit, or $\mathscr{L} = E_M/L$. The value of luminosity depends, in general, on the same factors as those determining the effective resolving power $\mathscr{R}$ and hence is not independent of the latter.

Variation of $\mathscr{L}$ and $\mathscr{R}$ with slit width

From the definition chosen for luminosity it is clear that the curve expressing the variation of E_M/E_0 also represents the variation of the ratio $\mathscr{L}/\mathscr{L}_0$ of the luminosity of a spectrograph having a slit width l to the luminosity $\mathscr{L}_0$ given by an infinitely wide slit.

The effective resolving power $\mathscr{R}$ of the instrument varies inversely as the image width. The inverse of the ratio d/d_0 is therefore equal to the ratio $\mathscr{R}/\mathscr{R}_0$ of the effective resolving power of a spectrograph having a slit width l to the intrinsic resolving power corresponding to an infinitely narrow slit. The influence of the slit width on the resolving power and luminosity of a spectrograph can be clearly seen in Fig. 2.8.

Bearing in mind that when the normalized slit width l/d_0 increases above 1·5–2 the intensity stays practically constant while the resolving power is a rapidly decreasing function of l, it is clear that there is no point in using a normalized slit width greater than 1·5 or 2. On the other hand the intrinsic resolving power $\mathscr{R}_0$ is approached as the slit is narrowed, but only at the price of a considerable loss of luminosity. It may therefore be expected that there is an optimum slit width that represents the best compromise between resolving power and luminosity, except in some exceptional cases. Since these two quantities vary in opposite directions the optimum is given by the value of l/d_0 that maximizes the product $(\mathscr{L}/\mathscr{L}_0)(\mathscr{R}/\mathscr{R}_0)$; this result is given approximately by $l/d_0 = 1$. We thus have $\mathscr{R} = 0.78\mathscr{R}_0$ and $\mathscr{L} = 0.82\mathscr{L}_0$ which represents a reasonably good compromise between the conflicting factors.

Departure from this value of l/d_0 brings a greater loss in the relative value of one of the quantities $\mathscr{R}$ or $\mathscr{L}$ than is gained on the other. These results, which assume incoherent illumination of the slit, remain qualitatively true for all conditions of illumination: only the optimum value of the slit width varies. It has the value $2d_0$ for the case of perfect coherence and lies between d_0 and $2d_0$ if coherence is only partial. This critical value of l obviously varies, for a particular instrument, with wavelength, since $d_0 = f\lambda/a$. It is easily determined experimentally with reasonably good accuracy; it is only necessary to observe a line in the spectral image plane while progressively reducing the slit width. The critical value of l is reached at the moment when a rapid reduction in intensity of the line first becomes noticeable, the intensity having up to that point remained practically constant.

We must not, however, lose sight of the fact that these considerations are relevant only to a perfect instrument, that is, one in which image quality is reduced by diffraction alone. In real systems, aberrations sometimes outweigh diffraction and so, to an even greater extent, does the granular structure of the photographic plate. The latter factor in particular demands further attention.

INFLUENCE OF THE PHOTOGRAPHIC EMULSION

2.9. *Nature of the effect*

Because of their granular structure, together with the complex phenomenon of scattering in gelatine, photographic emulsions must always have a limited resolving power. It is easy to prove that this is so and to measure the corresponding limit of resolution by photographing a series of Foucault targets; these are patterns formed by parallel bars of equal width, alternately black and white. When the distance between two consecutive black bars in the image is below a limiting value g, the bars are no longer separated. This limiting distance g depends on the emulsion and on the method of development (nature of developer, temperature, development time).[2] g is the limit of resolution of the emulsion under the given conditions; this is often called the 'grain of the photographic plate', but the nomenclature is incorrect because g is always significantly greater than the mean dimension of the silver grains after development. Its value, for emulsions in current use, varies between about 10 and 50 μm.

The existence of this limit of resolution in emulsions has an obvious consequence from the point of view of their utilization in spectrography. Two spectral lines will not be separable after development unless their separation on the plate is greater than g. Equally, a line image of width less than g produces, after development, a photographic image of which the width is of the order of g. It is therefore never profitable to image on the plate a spectral line of width less than g, since the resolving power remains at a value corresponding to $d = g$. There is, indeed, a serious disadvantage in narrowing the image to this extent: given equal slit illumination in the two cases, an image of width less than g produces less blackening of the emulsion than does one of width equal to g. This is explained by the fact that in the first case the incident luminous flux is used to darken an area of emulsion greater than that illuminated by the bright band. The blackening thus depends, not on the flux per unit area, but on the total flux received, which means that the image width should be at least $d = g$.

2.10. *Focal length and aperture for grain-limited resolving power*[4]

In determining the optical parameters of a spectrograph we can first test the possibility of ensuring that the presence of the photographic plate will not reduce the maximum resolving power that the instrument is capable of attaining. The minimum width of a spectral line in the focal plane of the objective O being $d_0 = f\lambda/a$, the photographic plate cannot reduce resolving power if its limit of resolution g is not greater than d_0. This gives

$$g \le f\frac{\lambda}{a}$$

For a given dispersing element, the width of the emergent beam is a (as defined in § 2.2), since under normal conditions of use it is the dispersing element that acts as the aperture stop. The above condition thus determines the minimum value for the focal length of the focusing lens:

$$f_R = a\frac{g}{\lambda}$$

f_R is the focal length for grain-limited resolution; it is usually large. Taking for example $a = 100$ mm and $g = 20$ μm we get for $\lambda = 0.5$ μm, $f_R = 4$ m. Quite ordinary values of a and g thus lead to values of f_R of the order of several metres, a situation having certain disadvantages, particularly of unwieldy size and lack of rigidity.

An even more serious consequence is that for a focal length f_R the corresponding relative aperture a/f_R of the focusing lens is always very small. From the relation $a/f_R = \lambda/g$, we find that in the case instanced above $a/f_R = 1/40$. a/f_R is the aperture for grain-limited resolution (which will be abbreviated to 'resolving aperture' from now on).

Now, the relative aperture a/f of the focusing lens determines the luminosity of the spectrograph. We know already that, for a given slit width, the luminosity is a definite fraction of the limiting luminosity $\mathscr{L}_0$ obtained with a wide slit (Fig. 2.8). But

GENERAL PROPERTIES OF SPECTROGRAPHS

from equation (2.5), $\mathscr{L}_0 = \tau\Omega$ or, if the illuminated area of the focusing lens does not depart too far from a square of side a,

$$\mathscr{L}_0 \approx \tau \left(\frac{a}{f}\right)^2 \qquad (2.6)$$

Since the luminosity varies with the square of the effective relative aperture of the focusing lens, the resolving aperture a/f_R, being itself very small, can only result in an exceedingly low value of luminosity; this is a serious drawback.†

2.11. *Aperture greater than the resolving aperture*

A higher value of luminosity than that corresponding to the resolving aperture can only be achieved at the expense of a reduction of resolving power. If the aperture of the focusing lens is significantly greater than the resolving aperture and hence the focal length f less than f_R, it is the photographic emulsion rather than the effect of diffraction which limits the resolving power of the spectrograph. The slit width l should be chosen so that the width d of its image is equal to the limit of resolution g of the emulsion, the latter being in this case significantly greater than the diffraction width d_0 of the image. This corresponds, on the curves of Fig. 2.8, to an abscissa l/d_0 in the region well above 1.

The luminosity of the spectrograph is thus very nearly equal to the limiting luminosity $\mathscr{L}_0 = \tau(a/f)^2$ while the effective resolving power $\mathscr{R}$ is inversely proportional to d, that is, to g. We now have:

$$\mathscr{L} \approx \tau\Omega \approx \tau \left(\frac{a}{f}\right)^2 \qquad (2.7)$$

$$\mathscr{R} = \mathscr{R}_0 \frac{d_0}{g} = \mathscr{R}_0 \frac{f}{a} \cdot \frac{\lambda}{g} \qquad (2.8)$$

† Equations (2.5) and (2.6) are only valid when the relative aperture of the objective O is fairly small. The error introduced by using equation (2.6) is negligible so long as a/f is less than $\frac{1}{4}$ but rises to 10 per cent when $a/f = \frac{1}{2}$ and thereafter increases rapidly as the aperture is widened. These very useful equations thus yield a good approximation in the majority of cases, provided that it is understood that they are not valid for the largest apertures, when they only indicate the direction of variation of $\mathscr{L}_0$ with a/f.

It is important to notice that under these conditions *the luminosity only depends on the relative aperture of the focusing lens,* which therefore emerges as a primary characteristic of the spectrograph; from now on this will be referred to simply as the *aperture* of the instrument. The luminosity of a spectrograph is thus determined by its construction and cannot be adjusted by the user.

The foregoing equations show that the effective resolving power $\mathscr{R}$ and the luminosity $\mathscr{L}$ of a spectrograph vary as functions of the aperture a/f but in opposite senses. More precisely, the product $\mathscr{L}\mathscr{R}^2$ is independent of the aperture; this product depends essentially on the intrinsic resolving power $\mathscr{R}_0$ of the dispersing element since $\mathscr{L}\mathscr{R}^2 = \tau\mathscr{R}_0^2\lambda^2/g^2$. The luminosity $\mathscr{L}$ may also be written

$$\mathscr{L} = \tau \left(\frac{\mathscr{R}_0}{\mathscr{R}}\right)^2 \frac{\lambda^2}{g^2} \qquad (2.9)$$

$\mathscr{L}$ therefore increases with the ratio $\mathscr{R}_0/\mathscr{R}$ of the intrinsic resolving power to the effective resolving power: this indicates that for a spectrograph of high luminosity the dispersing element should have an intrinsic resolving power significantly higher than the effective resolving power at which the instrument will operate. This conclusion can be expressed in another way: the higher the luminosity relative to that corresponding to the resolving aperture, the smaller is $\mathscr{R}$ compared with $\mathscr{R}_0$.

2.12. *Conclusion*

The results of this study may be summarized in the following way. The use of an aperture approaching the resolving aperture, which entails an excessively low luminosity, is only justified when it is absolutely necessary to achieve an effective resolving power as close as possible to the intrinsic resolving power of the dispersing element. This will never be the case when the dispersing element is a prism because it is always possible to replace the prism by a grating with much greater intrinsic resolving power and which can therefore offer a considerably higher luminosity for the same effective resolving power. (It will be shown later that the reflectivity of modern gratings is of the same order of

GENERAL PROPERTIES OF SPECTROGRAPHS

magnitude as the transmission factor of prisms; the coefficient τ in the equation $\mathscr{L} = \tau\Omega$ is therefore comparable in the two cases and, as already noted, in practical cases it is the solid angle Ω, and hence the relative aperture of the focusing lens, that determines the luminosity.) Even when the dispersing element is a grating, it is unusual to open up to the resolving aperture because a very high effective resolving power can be obtained at a very much higher luminosity by means of interference spectrometry.

In practically all cases, then, the aperture of spectrographs is considerably larger than the resolving aperture. To all intents and purposes luminosity depends only on this aperture; the slit width is chosen in relation to the limit of resolution g of the photographic emulsion in such a way that the width of the image shall be practically equal to g.

The effective resolving power is then

$$\mathscr{R} = \mathscr{R}_0 \frac{\lambda f}{ga}$$

Table 2.1 below summarizes the essential points of the preceding discussion (see also the curves of Fig. 2.8).

TABLE 2.1. Resolving power, luminosity and slit width

Aperture less than resolving aperture (exceptional case)	Aperture well above resolving aperture (normal case)
$\mathscr{R}$ and $\mathscr{L}$ depend only on normalized slit width l/d_0.	Luminosity determined by instrument components $\mathscr{L} \approx \tau\Omega \approx \tau(a/f)^2$
Best compromise: $l/d_0 = 1$ $\mathscr{R} \approx 0.8\mathscr{R}_0$ $\mathscr{L} \approx 0.8\mathscr{L}_0$	Slit width such that $d = g$ $\mathscr{R}$ limited by emulsion $\mathscr{R} = \mathscr{R}_0 \dfrac{\lambda f}{ga}$

It must not be forgotten that in all the foregoing discussion it has been assumed that aberrations have been so well corrected that they have no influence. This condition is easily satisfied when the focusing lens aperture is small but much less easily so in instruments having large apertures. The condition that the spectral line widths must never exceed the limit of resolution of the appropriate photographic emulsions nevertheless remains the operative criterion.

Note. The foregoing conclusions only apply to spectrographs with an extended source. The case of point sources presents an entirely different problem, encountered particularly in stellar astronomy. Here the slit width must be chosen so that all the flux forming the star image in the telescope gets into the spectrograph. The rules for setting up the stellar spectrograph are therefore different from those arrived at here; the expression for the luminosity in particular has a totally different form.[3]

2.13. *Recent developments*

In the preceding paragraphs, the luminosity of a spectrograph has been defined in terms of the ratio E/L of the intensity in the plane of the photographic plate to the radiance of the entrance slit. This definition is based on the fact that the blackening of the emulsion is a function of E; it is in accord with the everyday use of these instruments.

Recent research[1] has, however, shown that this mode of operation is, from a theoretical point of view, far from being the best. To make the best use of an emulsion, it is necessary to take account, not merely of the optical density, but of the signal/noise ratio, as for any other detector. The signal/noise ratio is defined as the ratio $D/\delta D$ of the optical density to the mean square deviation δD of this quantity due to granularity of the emulsion. This research has shown that the best signal/noise ratio does not correspond to the maximum optical density.

It thus becomes apparent that the photographic plate is used very inefficiently in spectrographs, especially in recording line spectra. The greater part of the sensitive area is not used at all and only a very small proportion of lines is recorded at an optical density compatible with a good signal/noise ratio.

A new method, known as grille spectrography, has been invented by P. Bouchareine and P. Jacquinot[1] to improve the utilization of the photographic plate. The spectrograph slit is replaced by a coarse grating identical with that used in the grille spectrometer to be described later (Fig. 6.21). The plate then records as many images of the grating as there are lines in the spectrum. The photograph so obtained is of course not directly decipherable and has to be decoded to reveal the spectrum. Under these conditions the whole of the sensitive surface is brought into use and the mean illumination can be adjusted so as to take advantage of the higher signal/noise ratio attainable.

When the photographic plate is employed in this way it can be shown that it is no longer the illumination at a point (the luminosity) that determines the signal/noise ratio but the total incident luminous flux. The situation is in this case analogous to that encountered with spectrometers.

CHAPTER THREE

PRISM SPECTROGRAPHS

3.1. General

The employment of prisms in spectrographs is, for practical purposes, only justified when the relative aperture of the spectrograph is considerably greater than the resolving aperture. When so used, the luminosity is satisfactorily high though the resolving power is naturally mediocre since the intrinsic resolving power of the prisms is never high. Reducing the relative aperture to the resolving aperture value leads to a ridiculously low luminosity: the same effective resolving power with a much higher luminosity can be achieved by replacing the prism by a grating, because of the much higher intrinsic resolving power of the dispersing element.

If an *a priori* choice of the effective resolving power and the luminosity is made, the constructional parameters of the spectrograph are predetermined, since

$$\mathscr{L} = \tau\Omega = \tau(a/f)^2$$

$$\mathscr{R} = \mathscr{R}_0 f\lambda/ag$$

The value chosen for the luminosity fixes the angular aperture Ω of the focusing lens and hence the relative aperture a/f. The choice of $\mathscr{R}$ then determines $\mathscr{R}_0$ and hence the parameters of the prism, though these are, of course, subject to physical limitations.

3.2. Choice of prism

By differentiation of the classical equations for the prism, the angular dispersion of a prism is seen to be

PRISM SPECTROGRAPHS

$$\frac{d\theta}{d\lambda} = \frac{e}{a} \cdot \frac{dn}{d\lambda}$$

in which e is the width of the prism base.

From (2.1), its theoretical resolving power is therefore:

$$\mathcal{R}_0 = e \frac{dn}{d\lambda} \qquad (3.1)$$

In this equation it is assumed that the full aperture of the prism is illuminated. If this is not so—see Fig. 3.1(b)—the equation becomes

$$\mathcal{R}_0 = (e_1 - e_2) \frac{dn}{d\lambda}$$

but since the prism aperture should normally be completely filled, Fig. 3.1(a) represents the usual situation. If the beam traverses several prisms in succession, the base widths are added.

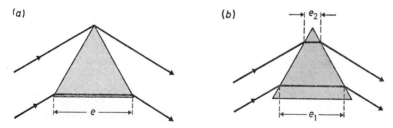

FIG. 3.1. Illustration of influence of method of illumination on intrinsic resolving power of a prism.

The expression $\mathcal{R}_0 = e\, dn/d\lambda$ shows that a spectrograph prism should be formed from a material that is both highly dispersive and very transparent in the relevant spectral region, the latter condition being necessary in order that the thickness may be as great as possible.

Choice of material

The two qualities of high dispersing power and high transparency are unfortunately not independent: indeed, they tend to be incompatible. As a consequence of a relationship between dispersion and absorption, optical materials have their highest dispersion near to absorption bands, just the regions in which they are useless as prism material. It follows that transparency takes first place in the criteria of choice of materials for use in the various spectral regions, but other considerations naturally have to be taken into account when a choice of materials is available. These are:

    Availability in sufficiently homogeneous quality
    Hardness (determining ease of polishing and durability)

For the present, the materials considered will be only those useful in the visible and ultra-violet regions: the present use of spectrographs is in fact restricted to these regions because of the insensitivity of photographic emulsions in the infra-red.† Only spectrometers are currently operated at wavelengths above about 1 μm; infra-red transparent materials will be dealt with when these instruments are described.

The choice of materials for the visible and ultra-violet is decidedly limited; flint glass and quartz (or fused silica) are practically the only ones available. At the short-wavelength end the limit of transmission of flint glass varies, according to its composition, from about 3500 Å to 4000 Å. Quartz is usable in the visible and ultra-violet down to about 2000 Å, making it a very commonly used material in spectrographs. If, on the other hand, only the visible spectrum is of interest, flint glass offers the advantage over quartz that its dispersion is about four times higher.

To be set against the outstanding qualities of quartz (exceptional durability, good homogeneity) there are nevertheless certain disadvantages. As with all natural crystals, it is sometimes difficult to procure sufficiently large specimens; moreover, it is birefringent and possesses optical activity—rotation of plane of polarization in propagation along the optical axis—so necessitating certain precautions in use. The crystals have in fact to be worked in such a way that the optical axis is perpendicular to the plane bisecting the prism apex. In addition, image doubling due to circularly polarized light has to be avoided; to

---

† An image converter tube could of course be substituted for the photographic plate but this solution is rarely adopted.

this end a Cornu prism (right-handed and left-handed quartz half-prisms cemented together) has to be made, or two separate prisms, one right-handed quartz and one left-handed, used in tandem.

For these reasons, quartz is progressively giving way to fused silica. For a long time the difficulty of getting large specimens of perfectly homogeneous silica stood in the way of its common use; this hindrance has now been removed and silica prisms are available with excellent homogeneity and high transparency in the ultra-violet.

A few materials, notably fluorite (crystalline CaF_2) and lithium fluoride (LiF), have a transparent range extending further into the ultra-violet than has quartz or silica. The transmission limit of fluorite lies in the region of 1250 Å, that of lithium fluoride is at about 1050 Å. Unfortunately, these materials have serious disadvantages; in particular, they are much softer than quartz and hence much more difficult to work optically and much more easily damaged. Until recently, their utility was also limited by the rarity of natural crystals of sufficiently large size and good quality. This is no longer a problem since artificial crystals of excellent quality can now be grown. Nevertheless they are not widely used because the manufacture of gratings has made such strides in recent years that they are preferred in place of prisms for spectrographs intended for studies in the far ultra-violet.

Choice of angle

As the intrinsic resolving power of a prism is proportional to the width e of the base, the apex angle of the prism should have the highest possible value. A limit is, however, imposed by the fact that as the angles of incidence and emergence increase with the apex angle, loss of light by reflection at the faces of the prism also increases; this effect is particularly serious when the prism is made of a high-index material.

The curves of Fig. 3.2 give the variation of the reflectivity of a boundary separating two transparent media of relative refractive index 1·65 as a function of angle of incidence. It is apparent that for angles of incidence greater than 60°, the loss of light by reflection rapidly becomes prohibitive. At 60° the corresponding apex angle is 63°, which is thus effectively its upper limit.

In practice, spectrograph prisms always have apex angles between about 50° and 70°. Fig. 3.2 also shows that in the region $i = 60°$, the reflectivity is different for the two directions of polarization, since this is close to the Brewster angle. It follows that emergent radiation from a prism is significantly polarized; this effect must be taken into account in certain applications.

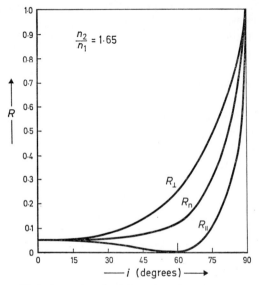

FIG. 3.2. Variation of reflectivity of a boundary between two transparent media (relative refractive index 1·65).
  $R_\perp$: electric vector perpendicular to plane of incidence.
  $R_\parallel$: electric vector parallel to plane of incidence.
  R_n: natural light.

Choice of dimensions

Limits are imposed on the dimensions of prisms either by absorption or by the size of available samples of good raw material. For crystalline materials such as quartz it is in fact the size of usable crystals that sets the limit, while in the case of glass or fused silica the difficulty of avoiding inhomogeneity increases rapidly with the volume of the sample. In certain very special cases this latter difficulty has sometimes been avoided by making use of a liquid prism but it is then essential to maintain a constant temperature for the prism to within very close limits.[5]

Recent progress in the manufacture of gratings has made it no longer necessary to strive at all costs for prisms of the highest resolving power; their performance will always remain far below that of a quite ordinary grating. The use of liquid prisms is therefore not worth considering. To obtain sufficiently high values of the base width e without involving extreme dimensions, two or three prisms are often used together in the form of a 'prism train'. The maximum values of e used in practice on commercial spectrographs are of the order of 300 mm, achieved, for example, by having a train of three prisms, each of 100 mm base width.

3.3. *Focusing lens*

Under normal conditions of use, the luminosity of a spectrograph only depends, in practice, on the relative aperture a/f of the focusing lens and it increases rapidly with the aperture; except in the case of very large apertures, the luminosity and relative aperture are connected by the relationship $\mathscr{L} \approx \tau(a/f)^2$, in which the transmission factor τ is not subject to significant variation. A high value of luminosity can therefore only be reached by having a focusing lens of high aperture; it is for this reason that, for instance, in the study of the Raman effect or night sky radiation, spectrographs having lenses of extremely high aperture, up to $f/0.6$, have been developed. However, the use of such lenses imposes serious problems if a reasonable value of resolving power is to be maintained. Thus, according to equation (2.8)

$$\mathscr{R} = \mathscr{R}_0 \frac{f\lambda}{ag}$$

it is necessary, in order to have a given effective resolving power, to use a dispersing element of which the intrinsic resolving power $\mathscr{R}_0$ is proportional to the relative aperture of the focusing lens. Now, the dimensions of a prism increase with $\mathscr{R}_0$ and so does the diameter of the lens, since the latter must not limit the beam width. Spectrographs of high luminosity must therefore be provided with focusing lenses of high relative aperture and at the same time of large diameter.

For the purpose of studying the light emitted by the night sky a spectrograph was developed twenty years ago with a focusing lens 250 mm in diameter and having an aperture $a = f/0.65$;[4] correction of the aberrations of such an objective for a wide spectral band presents a nice problem. This is of course an extreme example and in fact such a design project is no longer of practical interest since considerably better performance can be attained by other methods.

Apertures specified for commercial spectrographs range from $f/25$ to $f/1.5$ (with rare exceptions) for instruments having glass optics and from $f/25$ to $f/3.5$ for instruments with quartz optics. The availability of a wide choice of glass types of highly varied optical properties makes it possible to achieve better correction of aberrations in glass systems than in quartz; this accounts for the higher apertures available in the former systems.

In particular, it is impossible to eliminate the variation of spherical aberration with wavelength when only one optical material is available. It is interesting to note, however, that true chromatic aberration does not have to be corrected in a spectrograph: it is only necessary to tilt the photographic plate to the appropriate angle to bring all the spectral lines simultaneously into sharp focus.

A spherical mirror of large aperture has sometimes been used in place of the focusing lens, its spherical aberration being corrected by means of a Schmidt plate or a lens system. In the mounting due to Arnulf,[1] this correction is provided by the collimating lens.

3.4. *Collimator lens*

The focal length and f/number of the collimator lens have no influence on the fundamental characteristics of a spectrograph—that is, on its resolving power and luminosity. Since the usable diameter of this lens is approximately equal to that of the focusing lens, only one of these two quantities, either focal length or relative aperture, remains undetermined.

PRISM SPECTROGRAPHS

When the focusing lens has a very large aperture, a small relative aperture may be chosen for the collimator, with consequent advantages. Firstly, correction of aberrations is made easier; secondly, with a smaller collimator aperture it is easier to ensure correct conditions of illumination and, finally, since the focal length of the collimator will be greater than that of the focusing lens, the magnification γ of the system becomes less than unity. This is no small advantage because the slit width corresponding to a given image width is increased in the ratio $1/\gamma$, so making it possible to avoid the use of very narrow slits. Considerations of space and rigidity do, of course, impose a limit on the use of very large focal lengths for the collimator.

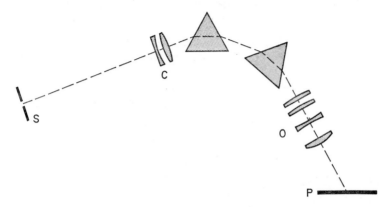

FIG. 3.3. Prism spectrograph. Classical mounting.

3.5. *The principal spectrograph mountings*

In the great majority of cases, the optical layout of prism spectrographs is based on that of Fig. 3.3, the collimator, prism (or train of prisms) and focusing lens simply being mounted in sequence.

The Littrow mounting, as shown in Fig. 3.4, is also sometimes employed though much less frequently with a prism than with a grating. This is in effect an autocollimator, utilizing the objective O as both collimator and focusing lens. The mirror m, which deflects the beam emerging from the slit, is actually placed just below the plane of the diagram and the spectrum is formed just above it so that m does not obstruct the return beam. The Littrow mounting has the advantage of compactness and, consequently, of increased rigidity compared with a classical mounting having lenses of the same focal length. There is, on the other hand, a defect of the system that is not easily surmounted: the objective O reflects an undesirable amount of radiation back towards the photographic plate. One expedient is to tilt the optical axis of this lens slightly, but astigmatism is then introduced. Carefully placed diaphragms are necessary to take out as much as possible of this unwanted illumination.

In Fig. 3.4, the prism angle is 30° and, since it is the rear face of the prism that acts as the mirror, the resolving power is that

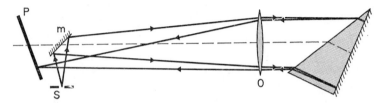

Fig. 3.4. Prism spectrograph. Littrow mounting.

of a 60° prism having the same face dimensions and mounted as in Fig. 3.3. Sometimes a 60° prism is used with a plane mirror behind it, giving the equivalent of a train of two 60° prisms.

3.6. *Examples of practical designs of prism spectrographs*

As has already been explained, commercial spectrographs are, almost without exception, equipped with focusing lenses having relative apertures between 1/25 and 1/1·5. The highest resolving powers naturally go with the lowest apertures and the instruments designed for these low apertures are generally intended for the spectrographic study of metals and alloys, especially steel. The spectrum of iron, which has an abundance of lines, is one that particularly calls for a relatively high resolving power, while the luminosity of the instrument can be low because the spark has a very high intensity.

PRISM SPECTROGRAPHS

Prism spectrographs of large aperture are generally designed for study of the Raman effect. Their resolving power cannot be high and is often inadequate; in such cases they can with advantage be replaced by a grating instrument.

Between these two extremes, all manufacturers produce instruments of medium characteristics for a variety of uses.

Details will now be given of three spectrographs representing each of the classes just described.

Low-luminosity instrument for spectral analysis of steel

Dispersing system: 2 quartz prisms with apex angle = 60°
Spectral region: from 2100 Å to 10 000 Å
Length of spectrum: 0·5 m
Focusing lens: aperture = $f/22$; focal length = 1·23 m
Collimator lens: focal length = 1·23 m
Inverse linear dispersion: $d\lambda/dx$ = 6·2 Å/mm at $\lambda \approx$ 3000 Å

Assuming a mean value of 20 μm for the 'grain' g of the photographic emulsion, the following values are readily calculable:

Spectral interval resolved in region λ = 3000 Å, $\Delta\lambda$ = $g\, d\lambda/dx$ = 0·12 Å
Effective resolving power of instrument $\mathscr{R}$ = $\lambda/\Delta\lambda$ = 25 000

The resolving power varies very rapidly with wavelength, in the same sense as the prism dispersion; it is about 40 times lower in the region of λ = 10 000 Å than at 2100 Å.

Ratio of effective resolving power of spectrograph to intrinsic resolving power of dispersing element:
The equation $\mathscr{R}/\mathscr{R}_0 = f\lambda/ag$ gives $\mathscr{R} = 0\cdot 33\mathscr{R}_0$.

Here we have a low-luminosity instrument, only suitable for very intense sources; on the other hand, its effective resolving power (25 000 at about 3000 Å, corresponding to a resolved spectral interval of 0·12 Å) enables this instrument to cope with present-day requirements for spectral analysis in metallurgy.

High-aperture instrument

Dispersing system: 2 flint prisms with apex angle = 63°
Spectral range: from 3900 Å to 8000 Å
Length of spectrum: 29 mm

Focusing lens: aperture $= f/1{\cdot}5$; focal length $f = 127$ mm
Inverse linear dispersion: $d\lambda/dx = 150$ Å/mm in region $\lambda = 5000$ Å

Again taking $g = 20$ μm, the following values are obtained at about $\lambda = 5000$ Å

Resolved spectral interval $\Delta\lambda = 3$ Å
Effective resolving power $\mathscr{R} = 1670$
Ratio $\mathscr{R}/\mathscr{R}_0 = 0{\cdot}037$

The serious loss of resolving power resulting from raising the relative aperture should be noted: the effective resolving power of the spectrograph amounts to less than 4 per cent of the intrinsic resolving power of the prisms themselves.

Medium-power instrument

Between the two preceding types of instrument are to be found numerous examples in which manufacturers have attempted to arrive at a compromise between average luminosity and an acceptable resolving power. Here is one example:

Dispersing system: 2 flint prisms of angle $61{\cdot}5°$, each having a base width of 70 mm
Spectral region: 3900 Å to 10 000 Å
Length of spectrum: 100 mm from 3900 Å to 8000 Å
Focusing lens: aperture $= f/8$, focal length $f = 400$ mm
Inverse linear dispersion: 45 Å/mm in region $\lambda = 5000$ Å

At about $\lambda = 5000$ Å and for $g = 20$ μm, the other data are:

Resolved spectral interval: $\Delta\lambda = 0{\cdot}90$ Å
Effective resolving power $\mathscr{R} = 5500$
Ratio $\mathscr{R}/\mathscr{R}_0 = 0{\cdot}2$

TABLE 3.1. Characteristics of typical prism spectrographs

Prisms	Aperture	Resolved spectral interval (Å)	Effective resolving power	$\mathscr{R}/\mathscr{R}_0$
Quartz	$f/22$	0·12	25 000	0·330 at about 3000 Å
Flint	$f/1{\cdot}5$	3	1 300	0·037 at about 5000 Å
Flint	$f/8$	0·90	5 500	0·20 at about 5000 Å

Table 3.1 demonstrates in no uncertain fashion the serious loss of resolving power resulting from any improvement of the luminosity of a spectrograph: for the largest aperture ($f/1\cdot5$) the effective resolving power is less than 4 per cent of the intrinsic resolving power of the prisms. This emphasizes the looseness of the connection between the effective resolving power of a dispersing instrument and the intrinsic resolving power of its dispersing element; only the former quantity is of practical value.

The most important conclusion from this study of prism spectrographs is that their performance can never exceed a relatively low level. The highest resolving power attainable does not go higher than 30 000 and this value is achieved only by deliberately reducing luminosity to an excessively low level. This naturally restricts the scope of the instrument very severely; conversely, if an instrument having a reasonable luminosity is called for, resolving power immediately falls to some very low value.

It is clear that prism spectrographs are inadequate to meet most of the requirements of spectroscopy; in the next chapter we shall see that grating spectrographs make it possible to reach an incomparably higher resolving power at equivalent luminosities, which suggests that an increasing use of this type of spectrograph may be expected.

CHAPTER FOUR

DIFFRACTION GRATINGS

PROPERTIES OF GRATINGS AND THEIR APPLICATION TO SPECTROSCOPY

4.1. *General*

A diffraction grating may be broadly defined as being a device capable of imposing a periodic deformation on the wavefront of any incident optical radiation. In practice, a grating consists of a series of rectilinear parallel, equidistant, rulings on a plane or spherical surface. It is these rulings that impress on the reflected or transmitted wavefront a deformation of spatial period proportional to that of the rulings.

A complete analysis of diffraction of light by a grating presents a very difficult problem. Various approximate methods have been explored but none is of general validity. In particular, it is not possible to ignore the vectorial character of electromagnetic waves without introducing errors which may be very significant.[3, 5, 17, 19, 27, 30] The only rigorously correct method consists in finding solutions to Maxwell's equations which satisfy the limiting conditions imposed by the surface form of the grating and by the properties of the grating material. This unfortunately is a problem of great complexity.[3, 21, 22] Although it would be impossible to deal here with the exact expression for the diffracted electromagnetic field, which depends on all the characteristics of the grating as well as on those of the incident radiation, we shall be able to discuss the general form of the diffracted field due to the periodic structure of the diffraction surface.†

4.2. *General form of periodic field diffracted by a grating*

The experimental arrangement of optical components used for the study of diffraction phenomena produced by a plane grating is shown diagrammatically in Fig. 4.1. Radiation emerges from a collimator composed of an objective C and a slit S parallel to the rulings of the grating G; after diffraction at

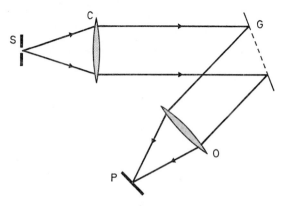

FIG. 4.1. Layout used for experimental study of diffraction by a plane grating.

the grating, it is collected by an objective O; observations are made in the focal plane of this lens.

The grating has a width l, measured perpendicularly to the rulings, and its pitch (distance between two successive rulings) is c. The length of the rulings is assumed to be infinite.

Taking a rectangular coordinate system $Oxyz$ (Fig. 4.2), let the plane xOy be the median plane of the grating surface, with the axis Oy parallel to the rulings. Assume initially that the slit S is reduced to a pinhole. The grating is then illuminated by a

† A complete and detailed study of diffraction gratings and their applications may be found in G. W. STROKE, 'Diffraction gratings', in *Handbuch der Physik*, vol. 29, 426–754.

plane monochromatic wavefront parallel to the rulings; the direction of the incident rays is **IO** in the plane of incidence $x0z$.

It has already been pointed out that the full theoretical treatment has to take into account the vectorial nature of the electromagnetic radiation field. In this present discussion it will be sufficient to consider a particular component of this field.

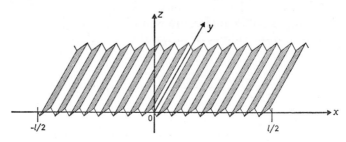

FIG. 4.2. System of coordinates used in the discussion of diffraction by a plane grating.

Let A_1 be the complex amplitude of one of the components of the incident field, and put $A_1 = 1$ in the wavefront (π)

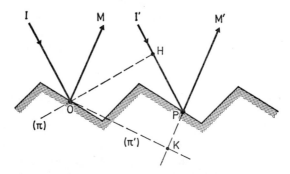

FIG. 4.3. Diffraction of a wavefront by a grating. Direction of the incident (**IO**) and diffracted (**OM**) rays.

passing through O (Fig. 4.3). At a point P (x, z) on the grating surface, this component is expressed by

$$A_1(P) = \exp\left\{-j\frac{2\pi}{\lambda}\overline{\text{HP}}\right\} \tag{4.1}$$

DIFFRACTION GRATINGS

in which $\overline{HP}$ denotes the distance of P from the wavefront, measured in the direction of the incident beam. This distance is $\alpha_1 x + \gamma_1 z$, α_1 and γ_1 being the direction cosines of **IO** on $0x$ and $0z$. If $z = h(x)$ represents the profile of the grating, we have

$$A_1(P) = \exp\left\{-j\frac{2\pi}{\lambda}[\alpha_1 x + \gamma_1 h(x)]\right\}$$

Now, $h(x)$ is a periodic function of which the period c is equal to the pitch of the grating. Hence

$$A_1(P) = a(x)\exp\left\{-j\frac{2\pi}{\lambda}\alpha_1 x\right\} = a(x)\,e^{-j2\pi u_1 x} \qquad (4.2)$$

$a(x)$ also being periodic, with period c.

Since the field at any point on the surface of the grating depends only on the abscissa x of that point, this surface may be divided into strips of width dx parallel to $0y$. Because of the periodic form of the surface the incident field (4.2) gives rise to a diffracted field in the direction **PM**′; the strip of grating having abscissa x produces a contribution to this field of which the complex amplitude may be written

$$d\mathscr{D} = b(x)\,e^{-j2\pi u_1 x}\,dx \qquad (4.3)$$

$b(x)$ having, like $a(x)$, the same periodicity as the grating. In the plane π' passing through O and perpendicular to the direction **PM**′ (Fig. 4.3), this field becomes

$$d\mathscr{D} = b(x)\,e^{-j2\pi u_1 x}\exp\left(j\frac{2\pi}{\lambda}\overline{KP}\,dx\right)$$

where $\overline{KP} = \alpha_2 x + \gamma_2 h(x)$, α_2 and γ_2 being the direction cosines of **PM**′ on $0x$ and $0z$. As before, this may be written

$$d\mathscr{D} = f(x)\,e^{j2\pi(u_2 - u_1)x}\,dx$$

where $f(x) = b(x)\exp\{j(2\pi/\lambda)\gamma_2 h(x)\}$ and $u_2 = \alpha_2/\lambda$.

The chosen component of the electromagnetic field diffracted in the direction **PM**′ by the whole grating of width l is therefore expressed by

$$\mathscr{D}(u) = \int_{-l/2}^{l/2} f(x)\,e^{j2\pi ux}\,dx \quad \text{(putting } u = u_2 - u_1\text{)}$$

This may conveniently be written in the form (see footnote, p. 15)

$$\mathscr{D}(u) = \int_{-\infty}^{+\infty} \text{rect}\left(\frac{x}{l}\right) f(x)\, e^{j2\pi ux}\, dx$$

$\mathscr{D}(u)$ is thus seen to be the Fourier transform of the function

$$\text{rect}\left(\frac{x}{l}\right) f(x)$$

All that we know of $f(x)$ is that it is a periodic function, of periodicity c, but this makes it possible to write it as a Fourier series

$$f(x) = \sum_{k=-\infty}^{+\infty} a_k\, e^{j2\pi kx/c}$$

whence

$$\mathscr{D}(u) = \int_{-\infty}^{+\infty} \text{rect}\left(\frac{x}{l}\right) \left[\sum_{k=-\infty}^{+\infty} a_k\, e^{j2\pi(u+k/c)x}\right] dx$$

$$= \sum_{k=-\infty}^{+\infty} a_k \int_{-\infty}^{+\infty} \text{rect}\left(\frac{x}{l}\right) e^{j2\pi(u+k/c)x}\, dx$$

If $G(u)$ represents the Fourier transform of $\text{rect}(x/l)$, we have

$$\mathscr{D}(u) = \sum_{k=-\infty}^{+\infty} a_k\, G(u+k/c) \qquad (4.4)$$

The Fourier transform of $\text{rect}(x/l)$ is well known; it is

$$G(u) = l\,\frac{\sin(\pi u l)}{\pi u l}$$

Hence

$$\mathscr{D}(u) = l \sum_{k=-\infty}^{+\infty} a_k\, \frac{\sin\{\pi(u+k/c)l\}}{\pi(u+k/c)l} \qquad (4.5)$$

The function $G(u)$ above is recognizable as the complex amplitude at infinity of a monochromatic radiation diffracted (in a direction given by the direction cosine $\alpha = \lambda u$) by a slit of width l illuminated by a wavefront parallel to its plane.

The curve (Fig. 4.4) of this function has a central section of which the width measured along the u-axis is $\Delta u = 2/l$, flanked by rapidly decreasing secondary maxima.

According to equation (4.4), the function $\mathscr{D}(u)$, representing the complex amplitude diffracted by the grating, is the sum of a series of functions identical with $G(u)$, each multiplied by its own coefficient a_k and centred on the abscissa $u = -k/c$. The central maxima of two successive functions $G(u+k/c)$ and $G[u+(k+1)/c]$ are therefore separated by an interval equal to $1/c$; this interval is very large compared with their width $\Delta u = 2/l$ since a practical grating always consists of a very large

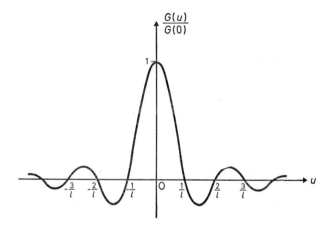

Fig. 4.4. $G(u) = \dfrac{l\sin(\pi ul)}{\pi ul}$, the Fourier, transform of $\mathrm{rect}(x/l)$.

number of rulings ($l \gg c$). It follows that $\mathscr{D}(u)$ can be regarded as being formed by the juxtaposition, without overlap, of a series of amplitude distributions identical with that shown in Fig. 4.4.

There is no difficulty in proceeding from the complex amplitude distribution $\mathscr{D}(u)$ to the corresponding distribution of intensity $I(u)$. Since the separate functions $G(u+k/c)$ are not superimposed, we have quite simply:

$$I(u) = \sum_{k=-\infty}^{+\infty} |a_k G(u+k/c)|^2 \qquad (4.6)$$

expressing the fact that it is only necessary to replace each of the amplitude distributions $a_k G(u+k/c)$ by the corresponding intensity distribution. A function $D(x)$ identical with $|G(u)|^2$ has already appeared in the discussion of the intrinsic resolving power of dispersing elements (§ 2.2, p. 9); its characteristic

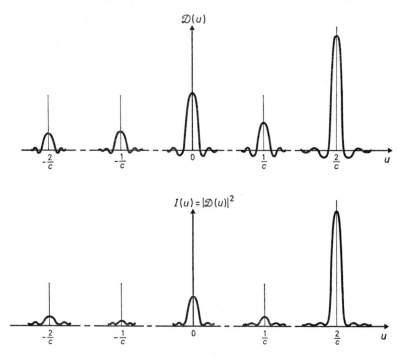

FIG. 4.5. Complex amplitude distribution $\mathscr{D}(u)$ and distribution intensity $I(u)$ of the wavefront at infinity diffracted by a grating. The incident wavefront is plane and monochromatic; here $u = (\alpha_2 - \alpha_1)\lambda$, α_1 and α_2 being the x-direction cosines of the vectors **IO** and **OM** respectively (Figs. 4.2 and 4.3).

curve is that of Fig. 2.3. It will be remembered that the width of this curve is determined by the difference of the abscissae of the central maximum and the first minimum, this width being also that of the main peak at the ordinate 0·405. In the present case this width is $1/l$.

When the collimator is provided with a slit parallel to the

grating rulings a series of bright lines is seen in the focal plane of the lens O (Fig. 4.1), separated by wide dark regions. Each very narrow line corresponds to one of the principal maxima of the function $I(u)$ and is bordered by a series of fringes of which the intensity falls off rapidly.

The directions corresponding to the principal maxima are given by the equation

$$u = -k/c = u_2 - u_1 = (\alpha_2 + \alpha_1)/\lambda \tag{4.7}$$

α_1 and α_2 standing for the direction cosines on Ox of the incident and diffracted rays **IO** and **OM** respectively.

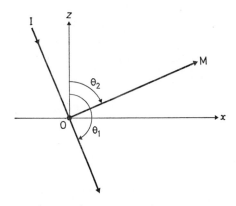

FIG. 4.6. Definition of angle of incidence θ_1 and of diffraction θ_2.

It is usual to express α_1 and α_2 as a function of the angle of incidence θ_1 and the angle of diffraction θ_2. Expressed algebraically (Fig. 4.6), we have:

$$\alpha_1 = -\sin\theta_1, \qquad \alpha_2 = -\sin\theta_2$$

and equation (4.7) becomes:

$$\sin\theta_2 - \sin\theta_1 = k\lambda/c \tag{4.8}$$

This is the classical grating equation, giving the directions of the principal maxima of intensity.

The angular width $\Delta\theta_2$ of the diffracted beams may also be calculated from the foregoing relationships. Expressed as a function of the parameter u, this width is $\Delta u = 1/l$, whence $\Delta \alpha_2 = \lambda/l$. Now, $\Delta \alpha_2 = \Delta \theta_2 \cos \theta_2$, from which $\Delta \theta_2 = \lambda/l \cos \theta_2 = \lambda/a$, a standing as usual for the cross-sectional width of the emergent beam (Fig. 2.2). As might be expected, this is the classical expression for the angular width of the diffraction pattern of a rectangular aperture already encountered in the calculation of the intrinsic resolving power of a dispersing element (see § 2.4).

Discussion of equation (4.8)

We have so far allowed the number k to take any integral value between $-\infty$ and $+\infty$. If, however, angle θ_2 is restricted to real values, equation (4.8) shows that the range of k lies between $-c(1 + \sin \theta_1)/\lambda$ and $c(1 - \sin \theta_1)/\lambda$. The number of values that k can assume is therefore limited and is also the number of principal maxima in the diffraction pattern generated by the grating. A more rigorous theory goes further, allowing k to take any integral value; outside the limits $-c(1 + \sin \theta_1)/\lambda$, $c(1 - \sin \theta_1)/\lambda$ the values of k are associated with complex angles of diffraction. The corresponding diffracted waves may be regarded as existing in the form of evanescent waves. In studying the practical applications of diffraction gratings to spectroscopy there is no need to take account of evanescent wave effects; the values of k will therefore be considered from now onwards to be contained within the limits defined above.

4.3. *Utilization of gratings in spectroscopy: intrinsic resolving power*

According to equation (4.8), the directions corresponding to principal maxima are dependent on wavelength: it is, of course, this property of gratings that accounts for their utility in spectroscopy. If, instead of the monochromatic radiation so far assumed, heterochromatic radiation falls on the grating, the directions of the diffracted beams are determined by the constituent wavelengths. A series of spectra is thus formed, equal in number to the permitted number of values of k. The value $k = 0$ is of course excluded since it corresponds to simple reflection or transmission without dispersion. The number k is called the *order* of the spectrum.

The intrinsic resolving power $\mathscr{R}_0$ of a grating is easily calculated. From equation (4.8), the angular dispersion is

$$\frac{d\theta_2}{d\lambda} = \frac{k}{c \cos \theta_2}$$

so from equation (2.1)

$$\mathscr{R}_0 = a \frac{d\theta_2}{d\lambda} = \frac{ka}{c \cos \theta_2} = \frac{kl}{c} = kN \qquad (4.9)$$

N being the total number of rulings in the grating.

This expression does not, however, indicate clearly the practical limits of a grating from the point of view of resolving power. The maximum usable order does in fact depend on N. The equation for $\mathscr{R}_0$ can be put into another form by using equation (4.8), from which

$$k = c(\sin \theta_2 - \sin \theta_1)/\lambda$$

Hence

$$\mathscr{R}_0 = cN(\sin \theta_2 - \sin \theta_1)/\lambda = l(\sin \theta_2 - \sin \theta_1)/\lambda \qquad (4.10)$$

l being, as before, the width of the ruled area of the grating.

From this expression the absolute maximum value of the resolving power is easily seen to be $\mathscr{R}_M = 2l/\lambda$. $\mathscr{R}_M$ is the *theoretical* limit of resolving power for a grating of width l with grazing angles of incidence and diffraction. The *practical* maximum resolving power of a real grating is always less than $\mathscr{R}_M$, the main reason being inevitable imperfections in the process of ruling; the higher the quality of a grating, the closer does its maximum resolving power approach the theoretical limit of resolving power. We are, of course, dealing here with the intrinsic resolving power of the grating and not with the effective resolving power of a dispersing instrument—spectrograph or spectrometer—in which the grating may be mounted.

It is worth noting that the theoretical limit of resolving power of a grating at a given wavelength is basically determined by its width alone. This means that the number of rulings per unit length is irrelevant, a conclusion that seems paradoxical when confronted with the well-known fact that the principal difficulties in making a grating arise from the requirement for a very large number (often a hundred thousand) of very close and regular rulings. We shall see that the very fine pitch is necessary to reduce to a minimum the overlap of successive orders of diffraction.

Nevertheless, gratings having only a small number of rulings of relatively wide spacing have been made (Michelson échelon) and some are still used (Harrison's échelle grating[9, 13, 29, 31]); their theoretical maximum resolving power is the same as that of a grating of the same width with closely spaced rulings.

4.4. *Superposition of spectra: free spectral range*

Any one direction of diffraction will correspond to that of the principal maxima of diffraction for several different monochromatic radiations, of which the wavelengths are:

$$\lambda = \frac{c}{k}(\sin\theta_2 - \sin\theta_1)$$

There are therefore as many possible values of λ as there are values of the order k. The wavenumbers of the superimposed radiations are:

$$\sigma = \frac{k}{c(\sin\theta_2 - \sin\theta_1)}$$
$$= |k|\sigma_1$$

with

$$\sigma_1 = \frac{1}{c|\sin\theta_2 - \sin\theta_1|} \qquad (4.11)$$

They thus form an arithmetic progression of which the common difference $\Delta\sigma_0$ is equal to the first term σ_1. If overlapping of spectra is to be prevented (and this is normally the case) then the range of wavenumbers corresponding to the spectral content of the incident beam must fall within an interval not exceeding $\Delta\sigma_0$. This property of $\Delta\sigma_0$ earns it the name of *free spectral range*.

It will be observed that $\Delta\sigma_0$ is equal to the inverse of the path difference δ between the rays diffracted in the defined direction

DIFFRACTION GRATINGS 51

by two successive rulings. (This relationship $\Delta\sigma_0 = 1/\delta$ is in fact quite general and is encountered in all interference systems.) It follows that the free spectral range varies inversely with the grating pitch c. The advantage of a fine-pitch grating is now apparent in its possession of a large spectral range without overlapping of orders.

Suppose, for example, that the complete visible spectrum between $\lambda_1 = 4000$ Å and $\lambda_2 = 7000$ Å is to be photographed without overlap. The free spectral range has to be $\Delta\sigma_0 = 1/\lambda_1 - 1/\lambda_2 \approx 10\,000$ cm^{-1}; hence $\sigma_1 = 1/c|\sin\theta_2 - \sin\theta_1| = 10\,000$ cm^{-1}. Suppose the Littrow grating mounting is to be used, with $-\theta_1 = \theta_2 = 30°$; then $\sigma_1 = 1/c$, so that $c = 10^{-4}$ cm $= 1$ μm. The performance specification therefore calls for a grating with the extremely fine pitch of 1 μm.

4.5. Importance of ruling profile: blaze effect

Equation (4.8) also shows that for a particular wavelength λ there can be several principal maxima corresponding to the various possible values of k. It follows that the incident energy is distributed between the spectra of different orders: this is undesirable when the grating is to be used in spectroscopy because as much as possible of the available energy is wanted in a single order.

The concentration of energy in a single spectrum can be achieved by a proper choice of the profile of the rulings. It was shown in § 4.2 that the intensity of the principal maximum of order k is $|a_k|^2$, a_k being the coefficient of order k of the Fourier series expansion of the function $f(x)$ that represents the periodic disturbance imposed on the incident wavefront by the grating. This intensity must therefore depend on the profile of the grating, but to forecast the distribution of energy between spectra of different orders the exact expression for the diffracted electromagnetic field has to be known. It is only quite recently that this very complicated calculation has been successfully carried out for certain cases.[22]

The lack of a rigorous method has enforced the use of approximate theories which are valid for practical purposes so long as the grating pitch is considerably greater than the wavelength but which lose their validity when c and λ are of the same order of magnitude. For reflection gratings the échelette profile is commonly used; in this the cross-section of the groove profile is triangular, the apex angle being usually about 90° (Fig. 4.7). If the incident rays are normal to the facet AB, there will be a wavelength λ_k for which, in order k, the diffracted rays return in the same direction. From equation (4.8), λ_k is given by

$$2 \sin \phi = \frac{k \lambda_k}{c} \qquad (4.12)$$

ϕ being the angle between the facets and the mean plane of the grating. Under these conditions it is found that, in general, the greater part of the diffracted energy associated with a radiation

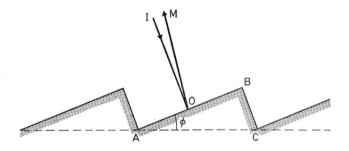

FIG. 4.7. Profile of an échelette grating.

of wavelength λ_k appears in the kth order, its direction corresponding to that of a specular reflection from the facets AB.

The concentration of diffracted energy in a single order only occurs, in principle, for rays of wavelength λ_k that are diffracted in the direction corresponding to specular reflection at the facets; in practice the effect is observable in a lesser degree over a fairly wide spectral band centred on λ_k.

This phenomenon is known as the 'blaze effect' and the angle ϕ is the blaze angle. It is only during recent years that the technique of making blazed rulings for the visible and ultraviolet regions has been available. Blazed gratings have given spectroscopic instruments a new power: the resulting gain in luminosity has made it possible to tackle problems in spectroscopy with instruments which were hitherto quite inadequate.

DEFECTS OF GRATINGS

The preceding analysis of the properties of gratings was based on the concept of an ideal grating perfectly ruled on a geometrically plane surface. Real gratings always show defects in a greater or lesser degree, especially defects due to irregularities in the ruling process resulting in the grooves not being straight, parallel and equally spaced; the correct profile is also difficult to reproduce. The consequences are, on the one hand, deformation of the diffracted wavefronts leading to deterioration of the images given by the grating or to the appearance of parasitic lines and, on the other, a loss of efficiency.

4.6. *Influence of defects on the diffracted wavefront*

The grating G (Fig. 4.8), of width l and pitch c would, if perfect, diffract plane wavefronts of width $a = l \cos \theta_2$ in directions inclined at angles θ_2 as defined by equation (4.8). At infinity each of these plane wavefronts would produce a diffracted image of the slit having an energy distribution of the classical form

$$\left(\frac{\sin(\pi u l)}{\pi u l}\right)^2$$

centred on the direction θ_2.

We take one of these favoured directions as the $0z$ axis (Fig. 4.8) and assume now that the grating has acquired defects: the diffracted wavefront moving in the direction $0z$ is now no longer plane but shows deformations resulting from these defects.

Let $\Delta\phi(x, y)$ be the phase difference at each point in the plane $x0y$ between the wavefront diffracted by the grating G and the ideal plane wavefront, of amplitude A_0, that would be diffracted by the same grating if it were perfect. At any point in plane $x0y$ the complex amplitude of the radiation is

$$A(x,y) = A_0 \operatorname{rect}\left(\frac{x}{a}\right)\operatorname{rect}\left(\frac{y}{b}\right) e^{j\Delta\phi}$$

b representing the length of the rulings (Fig. 4.8).

This amplitude distribution $A(x, y)$ in plane $x0y$ gives rise to a diffracted radiation at infinity in the direction **OM** having direction cosines α and β on axes $0x$ and $0y$ and of which

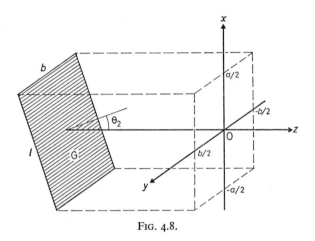

FIG. 4.8.

the complex amplitude $\mathscr{D}(\alpha, \beta)$ is the Fourier transform of $A(x, y)$.

$$\mathscr{D}(\alpha,\beta) = \iint A(x,y)\, e^{j2\pi(\alpha x + \beta y)/\lambda}\, dx\, dy$$

Hence

$$\mathscr{D}(\alpha,\beta) = A_0 \int_{-a/2}^{a/2} \int_{-b/2}^{b/2} e^{j\Delta\phi(x,y)}\, e^{j2\pi(\alpha x + \beta y)/\lambda}\, dx\, dy$$

Assuming that the quality of the grating is at least good enough for $\Delta\phi$ to remain small compared with unity, $e^{j\Delta\phi(x,y)}$ can be replaced by the first two terms of its expansion, so we have

$$\mathcal{D}(\alpha,\beta) \approx A_0 \int_{-a/2}^{a/2} \int_{-b/2}^{b/2} e^{j2\pi(\alpha x + \beta y)/\lambda} \, dx \, dy$$

$$+ jA_0 \int_{-a/2}^{a/2} \int_{-b/2}^{b/2} \Delta\phi(x,y) \, e^{j2\pi(\alpha x + \beta y)/\lambda} \, dx \, dy$$

$$= \mathcal{D}_0(\alpha,\beta) + \mathcal{D}_1(\alpha,\beta) \tag{4.13}$$

The first term $\mathcal{D}_0(\alpha,\beta)$ is simply the expression that pertains to a perfect grating: it represents the distribution of amplitude in the diffraction line produced by a perfectly regular grating. A real grating has the effect of superimposing an amplitude distribution arising from deformation of the perfect wavefront and represented by the second term $\mathcal{D}_1(\alpha,\beta)$. This is in fact one example of a more general statement of the effect of superimposing a perturbation on a perfect wavefront. Another example is to be found in an analysis of the method of image formation known as 'phase contrast': a wavefront representing a slight amplitude disturbance may be regarded as being composed of two superimposed wavefronts carrying radiation vectors in quadrature, one being the undeformed wavefront while in the other the radiation vectors have an amplitude proportional at each point to the deformation.

4.7. Periodic ruling errors: 'ghost' lines

The defects shown by gratings are of many different kinds; they arise either from the character of the surfaces on which the gratings are formed or from errors in the shape, parallelism and spacing of the rulings themselves. We shall only be concerned, for the present, with positional errors of ruling; these are the most important from the point of view of their effect and, equally, of the difficulty of eliminating them. The rulings may therefore be taken as being straight and parallel but showing a ruling error in the form of a displacement (measured by $\varepsilon(X)$ at any point having an abscissa $X = x/\cos\theta_2$) between the actual and the ideal positions of the ruling. It is easy to relate the deformations of the diffracted wavefront to the ruling errors which cause them. Thus, if the kth order of diffraction is being used, a phase difference $\phi = 2k\pi$ exists between the wavetrains diffracted by two neighbouring rulings of separation c. The error $\varepsilon(X)$ of one ruling therefore induces a phase shift $\Delta\phi = 2k\pi\varepsilon/c$ in the wavetrain diffracted by the ruling; the diffracted wavefront is thus deformed at the corresponding point by the amount $\lambda\Delta\phi/2\pi = k\lambda\varepsilon/c$.

The rulings were assumed to be parallel, so ε only depends on X. This function $\varepsilon(X)$, which of course reflects irregularities in the mechanism of the ruling engine, varies considerably from grating to grating. The most important form of this function is the periodic, and this we shall consider first. Ruling engines operate on the classical dividing engine principle, that is, a carriage carrying the grating and driven by a screw which may have a pitch of the order of a millimetre. It is extremely difficult to prevent periodic errors of a pitch equal to that of the screw from appearing in the movement of the carriage. The elimination of this error, which can have a disastrous effect on the performance of the grating, has for long presented a particularly obstinate problem to ruling engine designers; it is only quite recently that a really effective solution has been found.

We can examine the case of $\varepsilon(X)$ as a periodic function by writing it in the form of a Fourier series:

$$\varepsilon(X) = \sum_{n=-\infty}^{+\infty} b_n \, e^{j2\pi nX/P}$$

P being the spatial period of the function. Hence

$$\Delta\phi(x) = \frac{2k\pi}{c} \sum_{n=-\infty}^{+\infty} b_n \, e^{j2\pi nx/p} \quad (p = P\cos\theta_2)$$

and equation (4.13) becomes

$$\mathscr{D}_1(\alpha) = jA_0 \frac{2k\pi}{c} \int_{-a/2}^{+a/2} \left[\sum_{n=-\infty}^{+\infty} b_n \, e^{j2\pi nx/p} \right] e^{j2\pi\alpha x/\lambda} \, dx$$

which can be re-written as

$$\mathscr{D}_1(\alpha) = j\frac{2k\pi}{c} \sum_{n=-\infty}^{+\infty} b_n \left[A_0 \int_{-a/2}^{+a/2} e^{j2\pi\{(n/p)+(\alpha/\lambda)\}x} \, dx \right]$$

The quantity in square brackets represents the same distribution of complex amplitude as $\mathscr{D}_0(x)$ but centred on a direction defined by $\alpha = -n\lambda/p$. So

$$\mathscr{D}(\alpha) = \mathscr{D}_0(\alpha) + \mathscr{D}_1(\alpha) = \mathscr{D}_0(\alpha) + j\frac{2k\pi}{c}\sum_{n=-\infty}^{+\infty} b_n \mathscr{D}_0(\alpha + n\lambda/p) \quad (4.14)$$

From this result it can be seen that the diffraction line represented by $\mathscr{D}_0(\alpha)$ is accompanied by a series of parasitic lines having the same structure as the principal line and centred on the directions defined by $\alpha = n\lambda/p$, n taking all integral values, both positive and negative. These parasitic lines are called ghosts; they are arranged in symmetrical pairs on either side of the principal line at distances from the line which vary inversely with the pitch of the ruling error.

4.8. Intensity of ghosts

It is not difficult to predict the existence and position of ghosts without the help of calculations, since it is to be expected that a periodic ruling error would show its effect as a periodic distortion of the diffracted wavefront, as though the latter had itself undergone a second diffraction at a grating of pitch p in the plane xOy. For such a case, equation (4.8) clearly predicts the presence of amplitude maxima and minima located by direction cosines $\alpha = n\lambda/p$.

Equation (4.14) does, however, take us further by indicating the intensity of the ghosts; the ratio of intensity of the nth order ghost to that of the principal line is

$$I_n = \frac{4\pi^2 k^2}{c^2} |b_n|^2 \quad (4.15)$$

or, substituting for k from equation (4.8),

$$I_n = \frac{4\pi^2}{\lambda^2}(\sin\theta_2 - \sin\theta_1)^2 |b_n|^2 \quad (4.16)$$

Some points arising from these expressions are worth noting. Firstly, the symmetrical pairs of ghosts have the same intensity because, $\varepsilon(X)$ being a real function, $|b_n| = |b_{-n}|$. Secondly, the relative magnitudes of the ghosts depend on the values of the respective coefficients b_n and hence on the nature of the function $\varepsilon(X)$. Finally, the most important point is that the intensity of each ghost is proportional to the square of the spectral order or, put another way, to $(\sin \theta_2 - \sin \theta_1)^2$. Now, the resolving power of a grating is given by

$$\mathscr{R}_0 = l(\sin \theta_2 - \sin \theta_1)/\lambda = (\sin \theta_2 - \sin \theta_1)\mathscr{R}_M/2$$

so, from equation (4.16)

$$I_n = 16\pi^2 \left(\frac{\mathscr{R}_0}{\mathscr{R}_M}\right)^2 \frac{|b_n|^2}{\lambda^2} \qquad (4.17)$$

The suppression of ruling defects must therefore be pursued to the limit of practicability if the grating is to yield a resolving power as close as possible to the theoretical maximum, $\mathscr{R}_M$. In practice it is these defects that limit the usable resolving power.

Equation (4.17) also shows that the ruling error $\varepsilon(X)$ is associated with wavelength, since it only appears in the ratio $\varepsilon(X)/\lambda$; this means that the shorter the wavelength the more difficult it is to make a grating of a given performance.

The tolerances to be met are very tight; if, for instance, the total intensity of ghosts for a first order spectrum is not to exceed 1/1000 of the intensity of the principal line, we have the condition

$$\sum_n I_n = \frac{4\pi^2}{c^2} \sum_n |b_n|^2 = \frac{4\pi^2}{c} \overline{\varepsilon^2} < 10^{-3}$$

whence, approximately, $(\overline{\varepsilon^2})^{\frac{1}{2}} < 0.5 \times 10^{-2} c$. The root mean square of the ruling error $\varepsilon(X)$ must therefore not exceed 5/1000 of the grating pitch; in a 500 lines/mm grating this demands an average ruling error of less than 0·01 µm—a very small tolerance. This will give some idea of the difficulties to be faced in ruling good gratings.

Even with such close tolerances, ghosts can be obtrusive at higher orders: their combined intensity reaches one-tenth that of the principal line at the tenth order, so gratings for use at these high orders call for even higher precision of ruling.

In spite of these severe constraints, it is now possible, as we shall see, to reduce ghosts to a level at which they have negligible effect, even when resolving powers close to the theoretical limit are used.

4.9. Other defects: satellites, scattered radiation; importance of ruling profile

Apart from ghosts, other parasitic lines may be visible, appearing closer to the principal diffracted rays than the ghosts and having an irregular distribution. These are known as *satellites* and are due to non-periodic deformations of the diffracted wavefront produced by positional errors of the rulings or by imperfections of the surface on which the grating is formed. The intensity of satellites, as with ghosts, is proportional to $(\sin\theta_2 - \sin\theta_1)^2$ and so to $(\mathscr{R}/\mathscr{R}_M)^2$; they therefore also contribute to the limitation of the practicable maximum resolving power.

Random defects in the position and profile of rulings, as well as inevitable irregularities in their surface structure, give rise to scattered light which increases in intensity very rapidly as the wavelength is reduced. In the ultra-violet this glare can cause serious trouble and is very difficult to control, especially as gratings designed for this spectral region generally have a fine pitch (for example, 2000 to 3000 lines per millimetre). Nevertheless the reduction of scattered light to the absolute minimum is of the highest importance for the proper utilization of gratings, especially in spectrophotometry.

As might be expected, the profile of the ruled grooves never has exactly the intended form. This is an important point because the actual form directly influences the efficiency of the grating and the distribution of energy between the different orders of diffraction.

The échelette profile is produced by means of a diamond tool having its tip accurately shaped to an angle equal to the desired angle between two adjacent facets of the grating (Fig. 4.7); this angle is usually about 90°. The blaze angle is set by suitably inclining the diamond during the ruling. Even when the diamond is perfectly shaped, the achievement of a good profile calls for much care. The way in which the diamond does its work is not well understood; the metal is displaced rather than chiselled and this is reflected in the profile and finish of the surface so formed. It is also essential that the depth of cut be closely controlled: if it is too deep the furrows encroach on each other's space; if too shallow unwanted 'flats' remain between them. In either case the profile departs from the ideal and there may

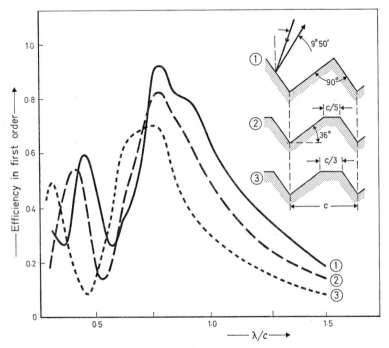

FIG. 4.9. Influence of profile flats on efficiency curves of an échelette grating. These curves represent the results of measurements in the millimetre wavelength region (after R. DELEUIL, *Optica Acta*, 1969, **16**, 23).

be a serious loss of efficiency. The curves in Fig. 4.9 give a comparison of the efficiencies of a perfect échelette grating and of two similar gratings having flats of different widths in the profile. In the latter examples it is seen that there is not only a lowering of the overall efficiency but also a general shift of the curves towards shorter wavelengths. The combination of these two effects can in certain cases result in an enormous gap between the efficiency of a properly ruled grating and one with grooves that are too shallow. It will now be appreciated that the control, to an adequate degree of precision, of the groove form is imperative.

4.10. *Inspection of grating quality*

The quality of gratings and the properties of their defects can be analysed in various ways. One of the most important assessments is that of the shape of the diffracted wavefront: techniques commonly used are those normally applied to the study

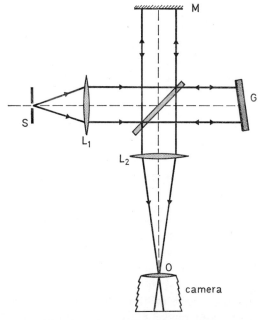

FIG. 4.10. Modification of a Michelson interferometer for studying defects in gratings.

of optical aberrations by measuring wavefront distortion, especially the interferometric method.[26, 31]

A Michelson interferometer is very suitable for this purpose, the grating to be examined replacing one of the interferometer mirrors (Fig. 4.10).

Monochromatic light illuminates a slit S and is collimated by lens L_1. The interferometer is adjusted so that the plane reference wavefront Σ_0 reflected by the mirror M is inclined at a small angle to the wavefront Σ_1 diffracted by the grating G; the inclination is in a direction such that the line of intersection of the wavefronts runs perpendicular to the direction of the rulings. This gives rise to wedge fringes on the surface of the grating perpendicular to the rulings. If the grating were perfect these fringes would be straight and parallel but in practice they show irregularities; at any point the displacement from the mean straight line represents the distortion of the wavefront Σ_1 relative to that of Σ_0. In particular, if this distortion is due to a ruling defect, it is easy to discover its magnitude. A deformation of the diffracted wavefront equal to $\lambda \Delta\phi/2\pi$ displaces an interference fringe by an amount $\Delta y = s \Delta\phi/2\pi$, s being the fringe separation. Since $\Delta\phi$ is related to the error ε in ruling position by the equation $\lambda \Delta\phi/2\pi = k\lambda\varepsilon/c$, it follows that the error ε produces a displacement of the fringe Δy such that $\Delta y/s = k\varepsilon/c$.

A straightforward measurement of the interference pattern thus gives information on the nature, position, and magnitude of the various defects of the grating.

Plate I reproduces an interferogram obtained with a 300 lines/mm grating set for the 4th order. It is quite easy to detect the presence of three significant defects. Near the left-hand edge of the photograph the fringes show an inverted V deformation and, further in, a sharp discontinuity; there is also a continuous waviness over their whole length. The nature of the ruling errors may thus be read off at a glance.

There are, of course, other methods which can be utilized to determine the shape of the diffracted wavefront:[1] among these strioscopy and phase contrast are worth considering.[4,14] The results are, however, not so readily interpreted as those obtained by interferometry; consequently these techniques are much less frequently used than the interferometry method.

A second problem, and one quite as important, is the accurate determination of the profile and surface finish given to the individual ruled grooves; we already know that these qualities have an important effect on the efficiency of the grating in each diffraction order and on the amount of randomly scattered light.

The electron microscope obviously has to be used for this purpose but the usual techniques of specimen preparation—replication of the surface followed by shadowing—are inadequate because they do not provide for the reproduction of the deep profile. Electron stereomicroscopy has also been tried but has not been particularly successful. New methods have therefore had to be developed, of which the most recent make it possible to observe both the groove profile and the surface structure at the same time, as Plates 2 and 3 demonstrate.

PRODUCTION OF GRATINGS

4.11. *Introduction and historical survey*[1,8,33]

Ruling engines are similar in principle to dividing engines; the grating is mounted on a carriage driven by a micrometer screw which, by means of a suitable mechanism, can be rotated intermittently through small equal angular increments so that the carriage moves forward step by step. During each stationary period a diamond rules one groove in a direction perpendicular to that of the carriage motion. It is also possible to make rulings on a continuously moving grating.[10]

The practical difficulties of manufacture arise, as might be expected, from the requirement of extreme precision in the finished grating—we have seen that the tolerance on the position of individual rulings is often less than one-hundredth of a micrometre. This being so, it is not surprising that the construction of a first-class ruling engine is an extremely difficult project.

Rowland was the first to lay down clear principles and he himself constructed three machines on which many high-quality gratings have been ruled. One of the key components is of course the screw, though it is not, as it happens, the most difficult to make. Rowland showed that it is possible to bring about a considerable improvement in the precision of a screw thread by running it in a long nut of the same pitch made up of several segments which are progressively tightened down onto the thread as it wears. There are, however, many other components of the engine that present some highly intricate problems of design and construction. For instance, the slightest eccentricity of the screw in its bearings produces a periodic error corresponding exactly to the pitch of the screw. Other considerations are the need to minimize the number of constraints imposed on the various components and to provide against effects of ageing such as wear and changes in elasticity of the materials, and so on. Last but not least is the problem of lubrication, accentuated by the fact that, at this extreme precision, any variation in the thickness of an oil film could cause unacceptable irregularities in the functioning of the engine.

It is quite remarkable that in spite of all these obstacles Rowland succeeded, as far back as 1885, in preparing gratings of excellent quality which achieved effective resolving powers of the order of 150 000. Moreover Rowland's ruling engines produced practically all the large gratings used throughout the world for half a century. With the same machines, reconditioned after various mishaps, Anderson carried Rowland's work further by producing resolving powers in the region of 350 000 while in 1915 Michelson achieved 400 000. These, however, were exceptional results and for a long period good gratings were very scarce. In fact, in spite of every effort since the time of Rowland and Michelson to improve the method of manufacture and, especially, to make it easier and more reproducible, many years passed before any substantial progress was made; this makes Rowland's achievements all the more outstanding.

In the last decade, however, the technique of ruling has made considerable progress; this has not only led to a significant improvement in the performance of gratings but has also rendered their quality much more consistent—a very important advantage.

4.12. *Recent progress in the technique of manufacture*

Gratings were originally ruled on the carefully polished surface of a solid block of metal: the metal was a specially formulated bronze (speculum metal). The hardness of this alloy and

DIFFRACTION GRATINGS 65

its lack of homogeneity had two undesirable consequences; one was the rapid rate of wear of the diamond, making it impossible to maintain the correct waveform. The second disadvantage was that harder grains in the speculum matrix deflected the diamond from its rectilinear course; the resulting irregularities caused an excessive amount of scattered light. The abandonment, in 1935 or thereabouts, of a massive metal blank in favour of a film of aluminium deposited by vacuum evaporation on a glass blank was therefore a major step forward. The aluminium film is much softer and much more homogeneous: it offers no hindrance to the ruling of perfectly straight grooves with a profile precisely controlled to take full advantage of the blaze effect. At the same time the creation of sources of scattered light is minimized.

As far as consistent repetition of groove spacing is concerned, the major advance has been the adoption of interferometric monitoring of the carriage movement.[2, 10, 11, 12, 31] It had long been obvious that only an interferometric system could provide the precise control of this movement with the precision demanded for a good grating, but before such a system could be put into practice certain other requirements had to be met. First of all, servomechanisms had to be available with a performance high enough to permit continuous correction of irregularities in the operation of the engine. Next, the demand for large gratings meant that path differences in the interferometer could be several decimetres: this required a light source capable of giving interference fringes of good contrast even at such large path differences. Recent progress in electronics and servomechanics coupled with advances in light sources of high monochromaticity such as pure isotope discharge tubes and, more recently, lasers have at last made the interferometric method practicable, especially in the hands of G. R. Harrison and G. W. Stroke, and with excellent results.

4.13. *Principle of ruling engines with interferometric control (Fig. 4.11)*

The carriage supporting the grating carries a mirror M which is part of a Michelson interferometer adjusted to give ring fringes at infinity; the other components of the interferometer are an integral part of the structure of the engine itself.

The central region of the ring fringe system is focused on a small circular diaphragm in front of a photomultiplier. The lamp must be a very narrow spectral line source; it may, for example, be a mercury-198 vapour lamp. As the carriage advances the interference fringes are developed in succession in front of the diaphragm which thus passes a sinusoidally modulated light flux; the current from the photomultiplier is then also a sinusoidal function of the position of the carriage.

To obtain a perfectly regular set of rulings, it is necessary to ensure strict synchronism between the progression of the carriage and the to-and-fro movement of the diamond stylus. For this purpose a second sinusoidal signal having nominally the same frequency as the first is generated to register the movement of the stylus. (This may be done, for example, by means of two polarizers, one fixed and the other rotated by the mechanism that drives the diamond; the pair of polarizers intercepts a light beam which is finally received by a photocell.) The condition for perfect synchronism between the movements of the carriage and the diamond is that the two sinusoidal signals shall remain constantly in phase. These two signals are therefore transmitted to a phase comparator, of which the output is a voltage proportional to the phase difference between the signals. This voltage represents an error signal: it is made to actuate a servomotor that acts, through a differential gear box, on the drive for the carriage screw. The rate of rotation of the screw is thus constantly adjusted to reduce the phase error between the two movement signals to zero. In this way the correct positioning of each groove is achieved to a very high degree of precision. In particular, the periodic ruling errors inherent in purely mechanical ruling engines are reduced to negligible proportions.

There are, naturally, many precautions to be taken to ensure proper functioning of the ruling engine. Very close control of temperature is essential. The machine is installed in an air-conditioned room in which the temperature variation does not exceed $0 \cdot 1$ °C, while the machine itself is immersed in an oil bath controlled to within $3-4 \times 10^{-3}$ °C. Variations in atmospheric

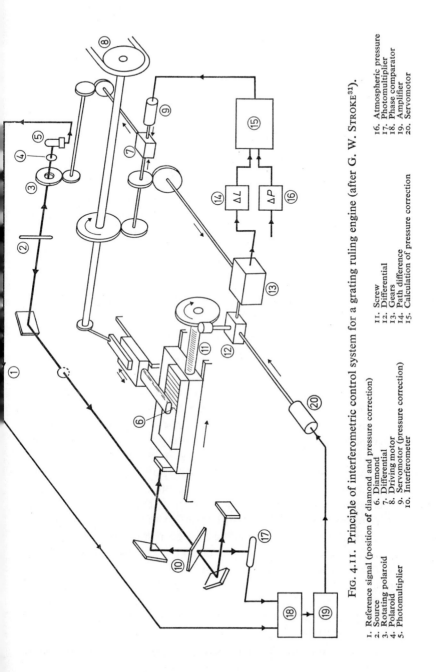

FIG. 4.11. Principle of interferometric control system for a grating ruling engine (after G. W. STROKE[31]).

1. Reference signal (position of diamond and pressure correction)
2. Source
3. Rotating polaroid
4. Polaroid
5. Photomultiplier
6. Diamond
7. Differential
8. Driving motor
9. Servomotor (pressure correction)
10. Interferometer
11. Screw
12. Differential
13. Gears
14. Path difference
15. Calculation of pressure correction
16. Atmospheric pressure
17. Photomultiplier
18. Phase comparator
19. Amplifier
20. Servomotor

pressure, which are certain to occur during the several days occupied by one ruling operation, must also be accurately compensated. The reason for this is that the path differences measured by the interferometer are a function of the wavelength of light in air and this varies with atmospheric pressure. A compensating system is therefore necessary to prevent these variations from being reflected in a change of ruling pitch.

4.14. Principal characteristics of gratings produced under interferometric control

The efforts to improve the quality of gratings are, of course, stimulated by the desire to achieve effective resolving powers as close as possible to the theoretical maximum value $\mathscr{R}_M = 2l/\lambda$; an increase in the usable width l is also valuable since $\mathscr{R}_M$ is proportional to l.

The resolving power of a grating set up with angles of incidence and diffraction θ_1 and θ_2, respectively, is in general

$$\mathscr{R}_0 = \tfrac{1}{2}(\sin\theta_2 - \sin\theta_1)\mathscr{R}_M$$

In order to take advantage of the blaze effect, a Littrow mounting or some similar type is normally used, for which $\theta_2 = -\theta_1$ (approximately) $= \phi$; in such cases, $\mathscr{R}_0 = \mathscr{R}_M \sin\phi$. It follows that, theoretically, the resolving power is increased simply by increasing the angle of incidence. We have, however, also seen that the magnitudes of the ghost and satellite lines increase, in general, with $\sin^2\phi$ so that in practice it is the grating defects that limit the effective resolving power actually available. As angle ϕ increases from zero, the effective resolving power increases at first, closely following $\sin\phi$, then passes through a maximum and finally falls off rapidly; the smaller the grating defects, the greater is the value of ϕ corresponding to the maximum effective resolving power. The gratings produced by Rowland and Anderson were never capable of being used effectively at angles above 10° or 20°; in contrast to these, values of ϕ up to 70° are practicable with the best modern gratings, yielding an effective resolving power exceeding 90 per cent of the limiting resolving power.[13, 28, 31]

In modern gratings the incidence of ghosts has been reduced,

DIFFRACTION GRATINGS 69

as we have already seen, to such a degree that their effect is quite negligible, at least in the visible and near ultra-violet regions. For example, a particular grating of 300 lines/mm ruled under interferometric control showed only two ghosts in the 11th order, each having an intensity equal to 1/3300 of the intensity of the principal line. These ghosts are completely negligible; they represent an extraordinarily small periodic ruling error. A measure of this error may be calculated from equation (4.15) which gives the intensity I_n of the nth order ghosts:

$$I_n = \frac{4\pi^2 k^2}{c^2} |b_n|^2$$

Since only two ghosts appeared, symmetrically disposed about the principal line, the ruling error $\varepsilon(X)$ must have been a sinusoidal function of amplitude, say, ε_M; under these conditions

$$b_1 = b_{-1} = \varepsilon_M/2 \quad \text{and} \quad I_1 = \pi^2 k^2 \varepsilon_M^2/c^2$$

Putting in the values provided above, we have:

$$\varepsilon_M/c = 5 \times 10^{-4}, \qquad \varepsilon_M = 1\cdot 7 \times 10^{-3} \, \mu\text{m}$$

The amplitude of the error thus is little more than a nanometre, a clear indication of the remarkable efficacy of interferometric control. Rowland ghosts are in fact no longer a problem except in the far ultra-violet, their persistence in that region being a natural consequence of the fact that their intensity varies as $1/\lambda^2$ [equation (4.17)].

Having practically eliminated ghosts, we are left with satellites as the real cause of the limitation of resolving power.[28] (Satellites, it will be remembered are parasitic lines occupying random positions in the immediate vicinity of the principal line.) These satellites are due to various aberrations of the diffracted wavefront generated either by defects in the flatness of the grating or by residual ruling errors. Satellites occur close to the principal lines because the wavefront deformations are of large period; their intensity increases, as for ghosts, with the angle ϕ and they therefore limit the effective resolving power in the same way as do ghosts. Satellites are particularly unwelcome when examining a weak line that has a strong line as a close neighbour.

The largest modern gratings are about 25 cm wide, ruled at 300–600 lines/mm, the length of a line being about 12 cm, giving a working area of 300 cm². At the green mercury line $\lambda = 5461$ Å the theoretical maximum resolving power of these gratings is

$$\mathscr{R}_\mathrm{M} = 2l/\lambda = 910\ 000$$

Effective resolving powers between 800 000 and 900 000 are reached in practice, a very creditable approach to the limiting value. As might be expected, performance is less good in the ultra-violet: at $\lambda = 2537$ Å the theoretical value for the same gratings is of the order of 2×10^6 but practical resolving powers do not exceed $1 \cdot 4 \times 10^6$. In the far ultra-violet the discrepancy is even greater.[13, 16]

Modern gratings are thus incomparably better than the Rowland or Anderson prototypes; they are capable of achieving resolving powers ten times higher than those original gratings, which means that their defects are one hundred times smaller. Plate 5 compares the interferograms obtained under identical conditions from an Anderson grating (upper interferogram) and a modern grating (lower).[28, 31]

Research aimed at further improvements in grating performance is now proceeding along two main lines. One follows the course of still further reducing the defects in order to increase effective resolving power in the ultra-violet while the objective of the other is to make wider gratings. The special problems associated with the production of gratings for the far ultra-violet will be discussed later.

Attempts have been made to rule a grating with two coupled diamonds, with the object of doubling the width to which gratings are normally limited when ruled by a single diamond.[25] This is, however, only profitable if the two halves of the grating diffract wavefronts that are perfectly in phase, which in turn imposes the condition that the separation of the two diamonds shall be an exact multiple of the grating pitch—and shall remain so throughout the ruling process. This method would therefore seem to be a very difficult one to apply. We may, nevertheless, hope for further advances in the performance of gratings in the not-too-distant future.

4.15. Production of replica gratings

Although the technique of ruling has now been brought to a high degree of perfection and the manufacture of master gratings of consistently high quality is possible, their production at a rate sufficient to meet the growing demands of spectroscopy is quite out of the question. Ruling engines are so difficult to construct and to maintain in good running order that their number is bound to remain very limited; to this must be added the further deterrent that to rule a 250 mm grating at the normal rate of ten lines per minute takes about ten days.

Fortunately it is now possible to take perfect replicas in quantity from a master ruled grating. The replicating material now used is a synthetic resin, which has replaced the collodion pellicle used in the earlier experiments. Collodion replicas were obtained by pouring a collodion solution over the master grating, allowing the solvent to evaporate, and then transferring the residual film onto a glass plate. Distortions occurring during these processes resulted in a replica of very poor quality.

The process employed at the present time may be briefly described thus: the master grating is first coated with a very thin film of a release agent to facilitate the eventual removal of the replica; an unpolymerized resin is then allowed to flow over the surface and is covered, while still fluid, with an optically flat glass plate. The resin polymerizes, causing it to harden and to adhere to the plate; the replica is then separated from the master and the moulded surface aluminized by the usual vacuum evaporation process.

The quality of good replica gratings is now so high that, judged by the results obtained, the replicas are indistinguishable from the originals; indeed some authors go so far as to suggest that the replicas are, in certain respects, superior to the master.

4.16. The Merton method of production[6, 7, 20]

In 1950, Sir Thomas Merton published a particularly novel method of ruling and reproducing gratings. The first stage of the operation consists in machining, on an ordinary lathe, a helical groove over one half of the length of a very accurately trued brass cylinder. The precision of this 'ruling' is naturally far below that necessary for a grating: the essential feature of Merton's idea lies in the method used to derive from this first helix a second, very accurate helix.

The standard method for making a lead screw of high quality is, as we have already seen, to work the screw in a nut of the same pitch, divided into segments, progressively tightening the nut onto the screw as unevennesses wear away until finally contact between the two is almost perfect as the nut runs along the whole length of the thread. (With a screw and nut of inferior quality movement is only possible when there is a certain amount of play between them.) Merton conceived the possibility of securing perfect contact between an imperfect screw and its nut without any preliminary 'running in' operation, simply by replacing the rigid nut by one having an elastic insert in which the screw would cut its own thread. On turning the screw, the nut, being in perfect contact with all the threads enclosed by it, must respond by a movement determined only by the mean pitch over the whole length of the nut; all irregularities within this length are absorbed by the elasticity of the insert.

The Merton nut takes the form of a metallic tube divided into three segments with the inner surface lined with a layer of suitable elastic material about one millimetre thick. For the first experiments cork was chosen but several other materials have been tried, among which elder pith seems to have given the best results. The procedure is, then, to cut as good a thread as possible over one half of a rod, as already described, then to clamp the Merton nut at the start of that thread, constraining it so that it cannot rotate but can move freely in the direction of the axis of the rod. On the nut is mounted a rigid bar carrying a diamond stylus which rules a second helix on the plain half of the rod when the latter is rotated: the pitch of the second thread is the same as that of the first but is much more regular.

Unfortunately the cylindrical ruling so produced cannot itself be used as a grating, so replicas have to be made from it by coating the cylinder with a film of plastic, slitting the film when set, removing it and mounting it on a glass plate. This first replica is used as a mould on which gelatine working gratings are formed.

Merton gratings cannot compete in quality with those produced by ruling under interferometric control; the reason is no doubt partly to be found in the complexity of the replicating process. The technique is, however, adequate for the manufacture of gratings to be used in infra-red spectroscopy and for educational purposes.

4.17. *Application of gratings in the far ultra-violet*[18]

Modern gratings, both original rulings and replicas, can be used without special precautions in the near ultra-violet because the reflectivity of the aluminium film on which they are ruled or with which the replicas are coated holds a value in the region of 90 per cent from the visible down to about 2000 Å. Below this wavelength the reflectivity falls steadily and is only about 15 per cent at 1000 Å. This deficiency at shorter wavelengths has encouraged the search for more efficient reflecting coatings.

It was soon discovered that the poor reflectivity of aluminium in the far ultra-violet is due, not to a property of the metal itself, but to the oxide film that normally forms on the metal in air; alumina has a very high coefficient of absorption for short wavelengths. To retain a high reflectivity the obvious precaution of protecting the metal against oxidization must be taken; this is done by depositing a transparent film on the aluminium coating immediately after the latter has been put down on the grating blank and without breaking the vacuum. By proper choice of its thickness the protective film can actually be made to enhance the reflectivity of the aluminium by inducing constructive interference within the wavelength range of highest importance. Even at the comparatively low pressure used in the aluminizing process some oxygen is present in the residual atmosphere: to prevent this affecting reflectivity, the aluminium has to be evaporated as rapidly as possible (1 to 2 seconds for a thickness of 800 Å) in the highest practicable vacuum; for the same reason the protective layer must be evaporated after the aluminium with the least possible delay.

Magnesium fluoride, transparent down to 1200 Å, is most commonly used for this purpose, the normal thickness being in the region of a few hundred ångströms. The curves of Fig. 4.12 compare the reflectivity of an unprotected aluminized mirror with reflectivities of similar mirrors protected with films of 250 Å and 380 Å thickness, all films being deposited under optimum conditions.

Below 1200 Å magnesium fluoride rapidly becomes absorbing but lithium fluoride is able to take its place as far as 1000 Å.

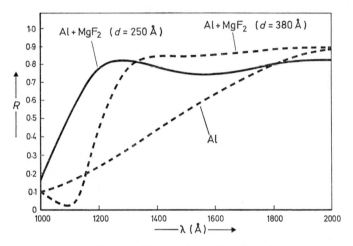

FIG. 4.12. Variation of reflectivity in the far ultra-violet of an unprotected aluminized mirror and of two aluminized mirrors coated with magnesium fluoride films 250 Å and 380 Å thick, respectively (after R. P. MADDEN, 'Preparation and measurement of reflecting coatings for the vacuum ultra-violet', in G. HASS, *Physics of Thin Films*, Academic Press, New York, 1963).

Thin films of this compound are not only soft but also hygroscopic; nevertheless, by protecting the lithium fluoride against water vapour with an exceedingly thin layer of magnesium fluoride a reflectivity of 60 per cent at 1000 Å has been attained.

These protected coatings may therefore be turned to good account not only for making reflectors but also for improving the efficiency of gratings at wavelengths extending down to 1000 Å. In the case of gratings, the ruled aluminium is, of course, already oxidized and must be given a second layer of high-reflection aluminium by the special processes just described.

Fig. 4.13 shows the increase in efficiency of a grating treated in this way.

Below 1000 Å, the coatings described are not effective and no other material has a particularly high reflection coefficient in this region of the spectrum. The best reflectors must be sought among the noble metals, which are not subject to surface deterioration. Platinum seems most likely to be useful, its

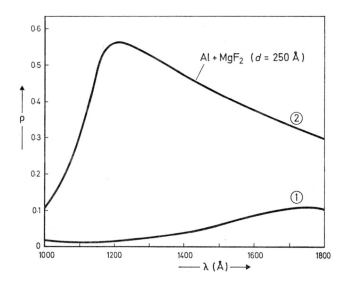

FIG. 4.13. Comparison between far ultra-violet efficiency of an ordinary grating (1) and that of a specially treated grating (2) (after R. P. MADDEN, 'Preparation and measurement of reflecting coatings for the vacuum ultra-violet', in G. HASS, *Physics of Thin Films*, Academic Press, New York, 1963).

reflectivity at normal incidence retaining a value of about 20 per cent between 500 Å and 1000 Å. Such poor reflectivities as this preclude the use of plane gratings because the corresponding spectrographs or spectrometers required two mirrors which, at short wavelengths, would impose prohibitive losses of intensity. Instruments with concave gratings are, however, feasible.

Another technique that can be used in this region and out to the lower limit of the ultra-violet spectrum is to employ a concave grating at grazing incidence. Theoretically the grating may be ruled on a material of any kind; the reflectivity of all materials closely approaches unity above a certain angle of incidence, usually higher than 80°. This well-known fact has often been exploited in rulings known as Siegbahn gratings on plain glass; the narrow grooves are shallow and are set at relatively large intervals. A high proportion of the incident radiation is, however, reflected by the flat surfaces between the rulings and represents wasted zero-order radiation, so that the efficiency of diffraction in the useful orders is very low. Modern technology now makes possible the preparation of metallic gratings for the extreme ultra-violet in which the profile of the grooves is adjusted so that the grooves themselves reflect at grazing incidence.[25] Various metals are used; unprotected aluminium proves to be an excellent reflector at grazing incidence (in spite of its oxidized surface), as do platinum and gold; this last offers a special advantage which often makes it the final choice: in thin-film form its structure is very fine-grained and hence its surface profile is very smooth. The result is that radiation scattered by the grating, which can so easily be an embarrassment in the far ultra-violet, is kept to a minimum.

4.18. *Manufacture of holographic gratings*

The idea of making gratings by recording the image of a system of interference fringes has long been recognized as offering a theoretically ideal solution to the problem of eliminating defects arising in the process of ruling.[34,35] The invention of the laser, with its high-intensity coherent beam, has transformed this idea into a practical possibility.

The first attempts to construct gratings in this way were made by recording on an ordinary photographic plate the interference fringes formed by superimposing two plane wavefronts produced by a laser. It proved to be possible to prepare large gratings in a few minutes but their poor quality and very low efficiency made them useless for spectroscopy.

The reason for this failure lay in the unsuitable properties of the emulsions normally used for photographic purposes. To avoid their use, A. Labeyrie developed a new photochemical process,[36] now successfully practised by the French firm Jobin et Yvon. A thin film of a photosensitive resin is deposited to a predetermined thickness on an optical flat true to $\lambda/4$. The coated surface is illuminated by two coherent beams, generated by an ionized argon laser, which thus give rise to a system of interference fringes. (During this exposure it is of course necessary to avoid any movement of the fringes by insulating the optical system from vibration, changes of temperature, atmospheric pressure, and so on.) When the exposure is terminated, the plate is immersed in a special solvent and the fringe system is developed as a grating in relief in the resin. The grating is aluminized *in vacuo* and finally surface coated for protection and optimum reflectivity in a particular spectral region.

These 'holographic' gratings, made by a purely static method, are obviously free from some of the defects inherent in gratings made by the traditional method and are not subject to the same limitations. It seems that, in particular, it should be possible to make gratings large enough for any practical purpose and with a very fine pitch; this is the type of grating specification which, as we have discovered, presents extreme difficulties when ruling is done mechanically.

The reported quality of the holographic gratings produced by A. Labeyrie, G. Pieuchard, J. Flamand and J. Cordelle is excellent;[37] examined with the Michelson interferometer the diffracted wavefront shows no appreciable defect and resolution tests indicate that the theoretical resolving power is reached. The amount of scattered light appears to be less than that for ruled gratings, probably because random errors in the line spacings are reduced.

The profile of the photographic grating lines is naturally different from that of ruled grooves. At the very fine line spacings for which holographic gratings are likely to be most useful (at least 1500–2000 lines/mm, with 6000 lines/mm or more as a possibility), the profile is practically sinusoidal; nevertheless experiment shows that the diffraction efficiency is of the same order of magnitude as that of échelette gratings, provided that the ratio of wavelength to pitch is greater than 0·3. For such small spacings the simple scalar theory, which indicates the triangular profile as producing the best blaze effect, is clearly not applicable. In the absence of a theory permitting the calculation of the ideal profile we can only accept the experimental evidence that a sinusoidal profile is effective.

The holographic method possesses a second advantage; this is the ability to produce stigmatic concave gratings. As we shall see, concave ruled gratings are highly astigmatic; Rowland himself recognized that it was theoretically possible to form stigmatic gratings, provided that the rulings could be made to conform to a particular pattern. More precisely, it is necessary that all the radiation emitted at a point A of the source and diffracted by a particular groove shall arrive in phase at a point A'. It is obviously impracticable to rule grooves mechanically in shapes that fulfil this condition; it is, on the contrary, quite possible to record interference fringes of the right shapes if the positions of the point sources of the interfering beams are correctly chosen. G. Pieuchard[38] has offered a simple analysis of the principle which is sufficient for an understanding of this method. Holograms formed in a similar way may also be designed to correct particular wavefront deformations, especially those due to aberrations in optical systems. Holographic correctors have, for example, taken the place of Schmidt plates for correcting spherical aberrations in cameras for the far ultraviolet, a region for which the making of Schmidt correctors is difficult or even impossible. These holographic correctors are produced by suitably deforming the wavefront of one of the beams forming the interference fringe pattern which constitutes the hologram.

CHAPTER FIVE

GRATING SPECTROGRAPHS

Grating spectrographs may be classified according to the type of grating for which they are designed; the two types are the plane grating and the concave grating. Although concave gratings enjoyed what almost amounted to a monopoly for many years, they are now much less in demand. The present ascendancy of plane gratings is due to a number of factors, of which the principal is that perfect control of the groove profile is more nearly attainable on a flat surface than on one that is concave. Another factor is that plane gratings are easier to incorporate in a satisfactory optical design, in particular one having a high relative aperture (and hence high luminosity) and no astigmatism. Concave gratings, on the other hand, still have one basic advantage, that of requiring no auxiliary optics to form a spectrum; this is of primary importance in the far ultraviolet below 1000 Å where, indeed, they have no rival.

PLANE GRATING SPECTROGRAPHS

5.1. General

Plane grating spectrographs are designed on the same lines as prism spectrographs and their general characteristics are correspondingly similar. The intrinsic resolving power $\mathscr{R}_0$ of gratings is nevertheless considerably higher than that of prisms (we found $\mathscr{R}_0$ reaching a few tens of thousands for prisms whereas the values for gratings can exceed a million). Thus the quantity

$$\mathscr{L}\mathscr{R}^2 = \tau(\lambda/g)^2 \mathscr{R}_0^2$$

(equation 2.9, p. 24) can attain much higher values for gratings than for prisms, so for equal luminosity grating spectrographs have an effective resolving power well above that of prism instruments and, inversely, they have a much greater luminosity for the same resolving power.

The operation of a prism spectrograph at an effective resolving power close to the intrinsic resolving power of the dispersing element is not, as we have seen, profitable because of the small relative aperture (and consequently low luminosity) that this condition imposes—it is always possible to have the same effective resolving power at a much higher luminosity by replacing the prism by a grating. The same argument cannot in turn be used to make a case for replacing the grating since no other dispersing element surpasses the grating in resolving power, with the exception of interference systems which are essentially different in their *modus operandi*.

It thus sometimes becomes necessary to push the performance of a grating to its limit, using it in a spectrograph operated at a relative aperture lower than the resolving aperture so that the grain of the emulsion cannot intervene as the factor limiting resolution.

5.2. *Description of the normal mounting*

The basic arrangement of all grating spectrographs is in principle that of the prism spectrograph shown in Fig. 2.1, but now the grating takes the place of the prism; also the collimator and focusing lenses are almost always replaced by mirrors. Full advantage is taken of the beam-folding effect of the mirrors by adopting mountings in which the incident and diffracted beams are kept close together, so making a compact and stable optical system. The most common layout is that shown in Fig. 5.1.

The relative aperture of the collimating mirror C has, by analogy with the lens in the prism spectrograph, no influence on the luminosity of the spectrograph and so may be made relatively small. The aberrations of this mirror are therefore negligible and do not need correction. This is not the case with the focusing mirror F: its relative aperture determines the luminosity of the instrument and may therefore have to be quite large.

In order to reduce off-axis aberrations it is usual to place the photographic plate as close as possible to the axis of the focusing mirror; this means, of course, that the plate holder must be as small as possible to reduce obscuration. For very large apertures

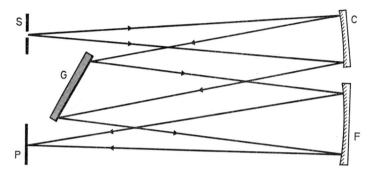

FIG. 5.1. Plane grating spectrograph: typical mounting.

it is essential to correct the spherical aberration of the mirror; correction is by means of a corrector plate (Schmidt aspheric corrector or a meniscus lens) which introduces an opposite aberration to that of the mirror[14] (Fig. 5.2).

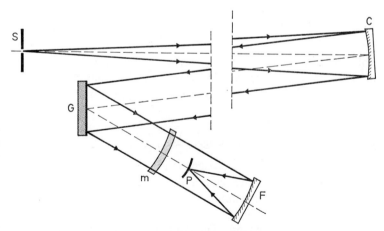

FIG. 5.2. Plane grating spectrograph using focusing mirror F of high aperture with fluorite meniscus corrector m (after C. J. SILVER-NAIL, *J. Opt. Soc. Amer.* 1957, **47**, 23).

5.3. Ebert mounting

Ebert conceived the idea of combining mirrors C and F in a single mirror M (Fig. 5.3).[4]

This mounting offers the advantages of rigidity and ease of adjustment. It can be shown that coma is corrected at the point on the photographic plate which is symmetrically opposite the slit with respect to the axis of the mirror (Z mounting), but this aberration increases rapidly away from this point. The Ebert mounting can be modified to reduce the inclination of the beams at the mirror by situating the slit above the grating and the photographic plate below it with the centre of the plate

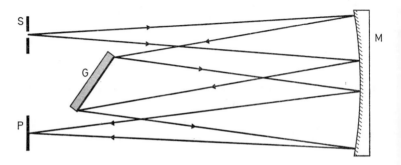

FIG. 5.3. Plane grating spectrograph: Ebert mounting.

on the mirror axis; the aberrations then remain within acceptable limits over the whole spectrum provided that the mirror aperture is not too large.[9] However, we shall see later that the Ebert mounting finds its most effective application in spectrometers (§ 6.14).

5.4. Multiple-camera spectrographs

The luminosity of a spectrograph is determined by the relative aperture of the focusing lens or mirror and this is fixed by the dimensions of the instrument. According to equation (2.8) the same is true for the effective resolving power, once the dispersing element has been chosen. The user of a conventional spectrograph has, therefore, no means of suiting the characteristics of the instrument to his immediate requirements by optimizing the enforced compromise between high resolving power and high luminosity.

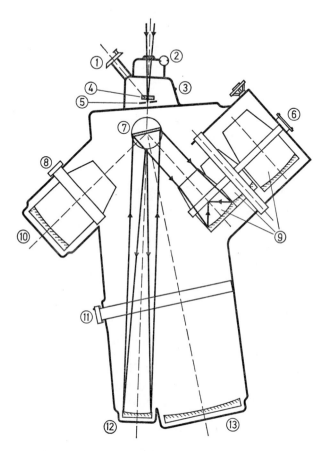

FIG. 5.4. Plane grating spectrograph equipped with five cameras (Haute-Provence Observatory) (after C. FEHRENBACH[5]).

1. Slit viewer
2. Field rotator
3. Calibration sources
4. Oscillating plate
5. Slit
6. Plate-changing port
7. Grating table
8. Plate carrier
9. Cameras I, II and III
10. Camera IV
11. Plate carrier
12. Collimator
13. Camera V

This disadvantage may be largely overcome by providing the spectrograph with several cameras of different focal lengths. Fig. 5.4 gives the layout of the spectrograph for the 1·93 m telescope at the Haute-Provence Observatory in Southern France;[5] it carries five cameras having the following dimensions:

Camera	Focal length (mm)	Aperture
I	165	$f/1 \cdot 1$
II } III	340	$f/2 \cdot 2$
IV	670	$f/4 \cdot 4$
V	2000	$f/13 \cdot 3$

The spherical aberrations of the camera mirrors I to IV are corrected by meniscus lenses. Camera V does not need a corrector plate. The collimator has a focal length of 4·5 m.

5.5. Order separators

When the bandwith of the radiation at the entrance slit is greater than the free spectral range of the instrument, there will be overlapping of spectra of different orders. To prevent this while still allowing an extended spectral range to be photographed in a single exposure, an order separator is sometimes used. This may consist of an auxiliary prism or grating dispersing system of which the dispersion is very much less than that of the main grating; the separator dispersion may be oriented so that the spectra of successive orders are formed one above the other on the photographic plate. The separator may alternatively be a small prism monochromator which forms a spectrum of low dispersion on the slit of the grating spectrograph itself, so restricting the incident spectral range; in this case the vertex edge of the prism is parallel to the spectrograph slit. Other examples of order separators will be found in the following sections.

5.6. Échelle grating spectrographs

Equation (4.10) showed that the resolving power of a grating does not depend on the number of grooves; very small pitch values are, nevertheless, employed in practice but they are the consequence of the need for a maximum free spectral range. If, however, an order separator is pressed into service, it becomes possible to make do with a reduced free spectral range and hence to utilize gratings with a coarser pitch. In this way a

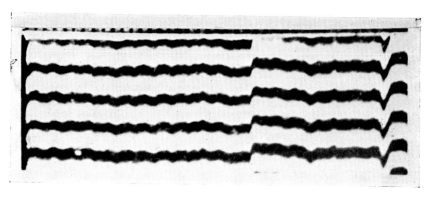

PLATE 1. Interference fringes obtained with the system shown in Fig. 4.10 (300 lines/mm grating, set for 4th order). (After G. W. STROKE, *J. Opt. Soc. Amer.*, 1955, **45**, 31.) (See p. 62.)

PLATE 2. Electron micrograph of profile of an échellette grating of 1230 lines/mm.† (See p. 63.)

† These micrographs were prepared in the Laboratoires d'Optique et de Physique Cristalline of the Faculté des Sciences de Marseille. For a description of the technique, see P. BOUSQUET, L. CAPELLA, A. FORNIER and J. GONELLA, *Appl. Opt.*, 1969, **8**, 1229. The grating was ruled, with interferometric control, by Société Jobin et Yvon (France).

PLATE 3. Electron micrograph of grating in Plate 2.† (See p. 63.)

† See footnote below Plate 2.

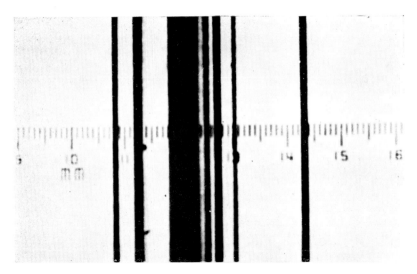

PLATE 4. Hyperfine structure of green line ($\lambda = 5461$ Å) of natural mercury (after G. W. STROKE[31]). (See p. 70.)

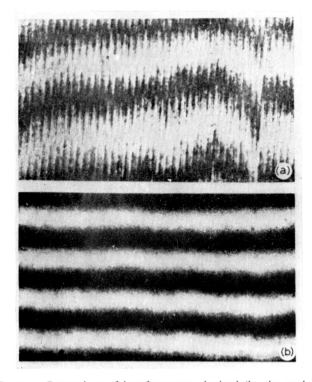

PLATE 5. Comparison of interferograms obtained (by the method shown in Fig. 4.10) from an early grating (*a*) (Anderson, 1914) and a modern grating ruled with interferometric control (*b*) (after G. W. STROKE[31]). (See p. 70.)

PLATE 6. Girard's hyperbolic grille. (See p. 127.)

PLATE 7. Fabry–Perot étalon with magnetostrictive scanning (after P. N. SLATER, H. T. BETZ and G. HENDERSON, *Proceedings of the Conference on Photographic and Spectroscopic Optics*, Tokyo (1964); *Jap. J. Appl. Phys.* 1965 **4**, Suppl I, 441). (See p. 169.)

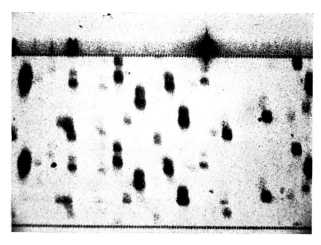
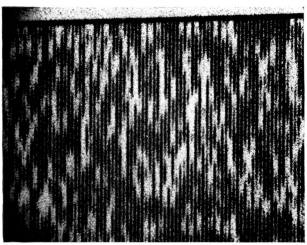

PLATE 8. Highly magnified reproductions of two sections of spectra obtained with the SIMAC (upper, emission spectrum; lower, absorption spectrum). The original spectra had a height of about 1 mm. On this magnified scale, their length would be about 1·4 metres.
(See p. 188.)

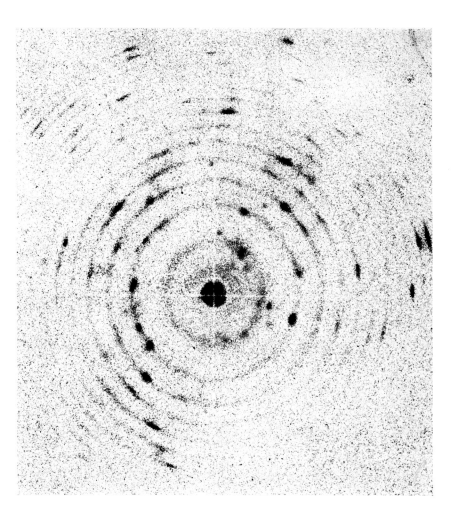

PLATE 9. Photograph taken with a mounting of the type shown in Fig. 7.25 by G. CARRANZA, G. COURTÈS, Y. GEORGELIN, G. MONNET and A. POURCELOT (*Ann. Astrophys.*, 1968, **31**, 63-100). (See p. 190.)

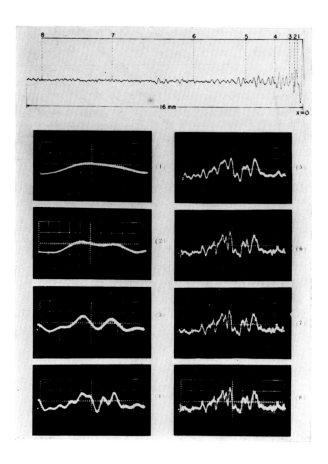

PLATE 10. Influence of the maximum value δ_M of path difference on resolving power. The top curve is the interferogram; oscillograms (1) to (8) show the spectra calculated from portions of the interferogram contained between $x = 0$ and the abscissae marked (1), (2), ..., (8) on the interferogram. The increase in the resolving power as δ increases is clearly shown (after H. YOSHINAGA, *Conference on Photographic and Spectroscopic Optics*, Tokyo (1964); *Jap. J. Appl. Phys.* 1965, **4**, Suppl. I, 426). (See p. 203.)

considerable gain in ease of manufacture may be looked for, especially in the control of groove profile.

These considerations led Harrison to the concept of the *échelle grating*; this has the normal triangular groove profile but a pitch of the order of tenths of a millimetre instead of a few microns.[6,7] The precision achievable under these conditions is so high that almost-grazing angles of incidence and diffraction can be used. The resolving power is then close to the theoretical maximum $\mathscr{R}_M = 2l/\lambda$ and the angular dispersion very high.

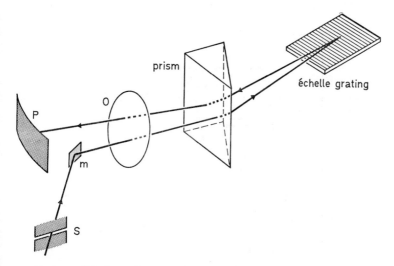

FIG. 5.5. Échelle grating spectrograph; the purpose of the prism is to separate the diffraction orders.

The low value of the free spectral range (about 50 cm^{-1}) invariably necessitates the employment of an order separator. A typical arrangement for a commercial spectrograph is shown diagrammatically in Fig. 5.5.† Separation of orders is provided by the quartz prism, oriented so that dispersion takes place in a horizontal plane; the slit, and hence the spectral lines, are horizontal as also are the échelle grooves, so that the grating dispersion is in the vertical direction.

† Reference (2) contains a description of another échelle grating spectrograph, in which orders are separated by means of a transmission grating.

Suppose that two radiations of wavenumbers σ_0 and σ_1 differ by an amount $\Delta\sigma_0$ equal to the free spectral range of the échelle and that the linear dispersion of the latter is such that the spectral lines A and B corresponding to σ_0 and σ_1 in the same order k (Fig. 5.6) are separated by a vertical distance slightly less than the vertical dimension of the photographic plate. These lines will also have a relative horizontal displacement due to their dispersion by the prism and this will be chosen to be slightly greater than their own length. The various spectral lines of order k occurring between A and B will therefore be centred

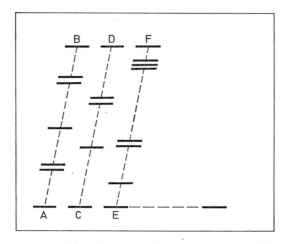

FIG. 5.6. Disposition of spectrum in the spectrograph of Fig. 5.5.

on the oblique line AB. Similarly, a spectral line C corresponding to $\sigma_1 = \sigma_0 + \Delta\sigma_0$ and of order $k+1$ is aligned horizontally with A; so also are other lines corresponding to $\sigma_p = \sigma_0 + p\,\Delta\sigma_0$, p being an integer. These different spectral lines are placed side by side, by virtue of the interposed prism, instead of being superimposed. Thus each segment AB, CD, EF and so on, displays a part of the spectrum contained within the free spectral range; each of these partial spectra belongs to a different diffraction order and the whole series can thus cover a very wide spectral range while at the same time having a comparatively high linear dispersion. One exposure can therefore record a great deal more information than is possible with the usual mounting because the area of the photographic plate is more fully utilized.†

5.7. *Multiple-pass and multiple-grating mountings*

To reach higher resolving powers and luminosities than can be attained with the largest available gratings by classical methods, various attempts have been made to use several gratings simultaneously or to return the diffracted beam to the grating once or several times. In the latter case, the mounting is classified as multiple-pass although in fact a double diffraction is usually regarded as the practical limit. It is easy to show that the angular dispersion $d\theta_2/d\lambda$ is then doubled and, since the width a of the beam is not altered, the intrinsic resolving power $\mathscr{R}_0 = a\, d\theta_2/d\lambda$ is also doubled.

Although a considerable advantage in dispersion and intrinsic resolving power has been gained in this way, it is only too obvious that the effect of grating defects is also enhanced; in particular the amount of scattered light is markedly increased and distortions of the diffracted wavefronts are doubled. Now that Rowland ghosts have been practically eliminated, the consequence of these distortions is, as we have seen, reduced to the presence of satellites in the immediate neighbourhood of the primary line; these satellites are naturally intensified by the double diffraction and the resolving power suffers, especially when two close lines of very different intensities are to be resolved. Nevertheless the gain in dispersion and resolving power offered by this type of mounting can be appreciable.[10]

In spite of the advantage of the double-pass mounting in extracting higher performance from a single grating, there is now a tendency to prefer a mounting long since adopted as standard for prism spectrographs, in which several dispersing elements are operative in series. This mounting is simpler and more compact than the multiple-pass mounting since no extra mirror is required; light losses are thus minimized. In addition,

---

† An analogous disposition of the spectrum occurs with the SIMAC interference spectrograph (see § 7.25).

careful selection of the two gratings brings about some compensation of their defects.

Fig. 5.7 is a diagram of an astronomical spectrograph constructed on this principle.[3] It would normally be preceded by a prism monochromator acting as an order separator.

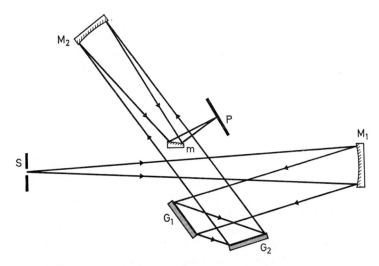

FIG. 5.7. Double plane grating spectrograph (after BARANNE[3]).

The apparatus outlined in Fig. 5.8 employs two Harrison échelle gratings E_1 and E_2 in series, while the concave grating G, with its rulings perpendicular to those of the échelles, separates the orders. The use of a concave grating reduces the number of reflecting surfaces to a minimum. [8, 15, 16]

If full advantage is not taken of the gain in resolving power that is available from these combinations of dispersing elements, an appreciable bonus in luminosity may often be drawn. The equation $\mathscr{L} = \tau(\lambda/g)^2(\mathscr{R}_0/\mathscr{R})^2$ indicates that for a given effective resolving power $\mathscr{R}$, doubling the intrinsic resolving power $\mathscr{R}_0$ would increase luminosity by a factor of 4 if the τ term remains constant. This would of course, only be true if τ were unity, which is never the case. The efficiency of gratings is, however, generally high enough to produce a significant gain in luminosity.

A secondary advantage is that for identical effective resolving powers the two-grating instrument is more compact than its single-grating equivalent: doubling the angular dispersion allows the focal length to be halved. The shorter focal length is, of course, the origin of the higher luminosity already noted. If,

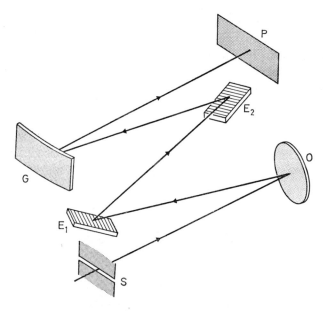

FIG. 5.8. Spectrograph using two échelle gratings, E_1 and E_2, and an order-sorting grating G (after G. R. HARRISON and G. W. STROKE, *J. Opt. Soc. Amer.* 1960, **50**, 1153).

however, the purpose is to reach higher resolving powers it is necessary to keep to the original focal length and to forgo the gain in luminosity.

Another way of using several gratings in series consists in arranging them side by side in the form of a mosaic, so simulating a grating of larger area. The positioning of each grating calls for great precision because, for a real gain in resolution, the diffracted wavefronts from each grating must be strictly in phase.

CONCAVE GRATING SPECTROGRAPHS

5.8. *Principles*

Rowland was the first to demonstrate that a grating ruled on a spherical concave surface can form a spectrum without the help of any auxiliary optics, itself combining the functions of dispersion and focusing. Although concave gratings are not as frequently employed now as they have been, their properties are still extremely valuable, especially in the far ultra-violet.

If P is the pole of the grating and C its centre of curvature (Fig. 5.9) the circle on the diameter PC (normal to the rulings)

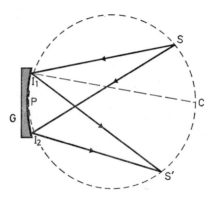

FIG. 5.9. Formation of spectrum by a concave grating.

is known as the Rowland circle. For the rays entering through the slit S to be correctly focused, the slit must be located on the Rowland circle and be parallel to the rulings. For each wavelength, the radiation from a point in the slit is concentrated, after diffraction, along a focal line (the tangential astigmatic focus) parallel to the rulings and passing through a point S', also on the Rowland circle. The spectrum is therefore formed along this circle, the spectral lines being formed by the alignment of the tangential focal lines corresponding to elementary lengths of the slit; the concave grating is therefore an astigmatic component.

GRATING SPECTROGRAPHS

Many mountings designed to take advantage of the unique property of the Rowland circle have been described, but not all have survived to the present day. The first of these, the Rowland mounting itself, is only worth a mention here because of its historical interest.

5.9. Paschen–Runge mounting

Among mountings in present use, one example is the Paschen–Runge, shown in Fig. 5.10. All the components (slit, grating, photographic plate) are at fixed points; a large number of plate holders therefore have to be located along the Rowland circle,

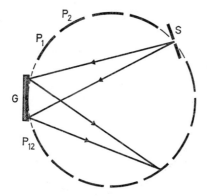

FIG. 5.10. Concave grating spectrograph: Paschen–Runge mounting.

which is represented by a large, very rigid, circular base. The principal advantage of the Paschen–Runge mounting lies in its ability to accommodate a large spectral interval in a single photographic exposure by virtue of the simultaneous use of several photographic plates. On the other hand, it is a very cumbersome mounting. Owing to its great stability, it is suitable for gratings of large radii of curvature (for example, 10 metres) and thus allows high resolutions to be attained.

5.10. Eagle mounting

In its basic form, the Eagle mounting is characterized by the coincidence in direction of the incident and diffracted beams corresponding to the mean wavelength radiation of the spectrum covered. From this point of view it is related to the Littrow mounting (§ 3.5, Fig. 3.4), particularly through the utilization of an auxiliary plane mirror to deflect the incident beam through 90° (Fig. 5.11).

In modern practice a variant of the Eagle mounting is favoured; in this the incident and diffracted beams are not quite coincident, the slit and photographic plate occupying

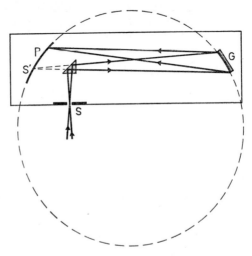

FIG. 5.11. Concave grating spectrograph: Eagle mounting.

neighbouring positions (Fig. 5.12). The advantages of the original Eagle mounting are preserved but the modified mounting gains by elimination of the auxiliary mirror; this mirror can only have very poor reflectivity in the far ultra-violet where concave gratings now find their only important application.

The Eagle mounting offers several advantages. Because the incident beams are not widely separated, astigmatism is slight and can easily be corrected by placing a weak cylindrical lens between the slit and the grating. Furthermore, the compact layout lends itself to a rigid construction and to operation under vacuum since the whole instrument is of a shape that can readily be enclosed in a tubular vacuum chamber.

An inconvenient feature of the Eagle mounting is the requirement for three independent adjustments when changing from one spectral region to another: the grating has to be moved along the line S'G and rotated about an axis through its pole parallel to the rulings, while the photographic plate has to be turned about an axis in its own plane. Instruments of low dispersion (with gratings of about one metre radius of curvature) do not, of course, suffer from this disability because the whole

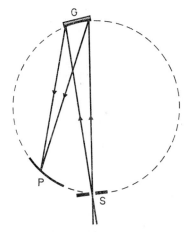

FIG. 5.12. Modified Eagle mounting.

spectrum can be contained within the width of the plate at one exposure; several commercial spectrographs make use of this or similar layouts.[11, 12, 13]

When the components have to be contained in a vacuum chamber it is often convenient to mount the slit and the photographic plate in positions not precisely on the Rowland circle. The slit will then be slightly outside the circle and the plate slightly inside, due care being taken to ensure that the conditions for focusing are satisfied. To complete this account of mountings, it is worth noting the off-plane Eagle mounting, in which the slit is above the Rowland circle and the plate below it, an arrangement analogous to that sometimes used for the Ebert spectrograph (see § 5.3).

5.11. Gratings used at grazing incidence and diffraction

The poor reflectivity of metals at normal angles of incidence in the far ultra-violet, especially below 500 Å, is a limitation that can be overcome, as has already been noted, by using the grating at grazing incidence. The mounting is then as indicated in Fig. 5.13, angle i being generally between 82° and 89°.[1]

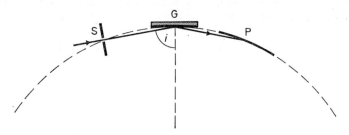

FIG. 5.13. Concave grating spectrograph for grazing incidence.

Under these conditions the linear dispersion is high: for example 0·5 mm/Å at $\lambda = 1000$ Å and 1·25 mm/Å at $\lambda = 100$ Å for a 600 lines per mm grating of 3 m radius. Unfortunately such high angles induce considerable astigmatism and this causes a serious loss of luminosity.

CHAPTER SIX

PRISM AND GRATING SPECTROMETERS AND MONOCHROMATORS

We now turn to the study of dispersing instruments that measure luminous flux rather than those that register images. The instruments now to be described are therefore *spectrometers*.

With the exception of a recent innovation (the *grille spectrometer*, to be described later), all the spectrometers in this chapter have two slits, S_1 and S_2 (Fig. 6.1), whence the designation *slit spectrometer*, currently used to distinguish these from interference spectrometers.

GENERAL PROPERTIES OF SLIT SPECTROMETERS

6.1. Description of the basic slit spectrometer

Prism and grating spectrometers differ from their spectrograph counterparts only in replacement of the photographic plate by a radiation detector, such as a photoelectric cell or a thermopile, capable of measuring the luminous flux directly. Excepting concave grating spectrometers, the basic layout of these instruments is that of Fig. 6.1, which may be compared with Fig. 2.1 for the case of the spectrograph.

In place of the photographic plate there is a second slit S_2, the exit slit; its function is to select from the spectrum formed by the dispersing element a spectral element, of width $\Delta\sigma$, characterized by a mean, or *setting*, wavenumber σ'. The flux Φ

contained within this elemental waveband is measured by the receiver R, which generates a signal proportional to Φ.

The spectrum is explored by continuously displacing the centre of the spectral band isolated by the exit slit. This is normally done by rotating the prism or grating and recording the varying signal from the receiver. This produces a curve representing the function $S(\sigma')$ which expresses the relative variations of flux at the exit slit as a function of the setting wavenumber σ'.

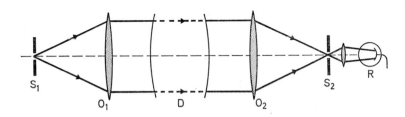

FIG. 6.1. Basic layout of a slit spectrometer.

The problem is then to deduce, from the recorded function $S(\sigma')$, the function $L_\sigma(\sigma)$ expressing the variation of the monochromatic source intensity L with wavenumber. $L_\sigma(\sigma)$ is simply the spectrum of the source, often called the source function: its precise determination is, of course, the ultimate aim of spectroscopical instrumentation.

$S(\sigma')$ and $L_\sigma(\sigma)$ are related by a linking function, the *instrumental profile*, now to be defined.

6.2. Definition of the instrumental profile of a spectrometer

Suppose the spectrometer to be receiving strictly monochromatic radiation. The curve recorded as the spectrum is scanned will not be an exact representation of the single spectral line but will have a finite width due to the effects of diffraction, the finite width of the slits, and other causes such as aberrations; in other words, the resolving power of the instrument is not infinite. The curve recorded under these conditions represents the instrumental profile. Put mathematically, if the energy distribution of the source as a function of wavenumber σ is a delta function $\delta(\sigma-\sigma_0)$ centred on the wavenumber σ_0, the recorded spectrum will be a curve representing a function $A(\sigma'-\sigma_0)$; then by definition $A(\sigma')$ is the instrumental profile of the spectrometer.

6.3. Relationship between the instrumental profile, the source function and the recorded function

It is obvious that the narrower the instrumental profile, the less the recorded curve $S(\sigma')$ departs from the true *spectral energy* distribution $L_\sigma(\sigma)$ of the source. The relationship between $S(\sigma')$, $L(\sigma)$ and $A(\sigma')$ is established by the following argument. Since the recorded function $A(\sigma'-\sigma_0)$ is produced by the spectrometer from a spectral line of zero width, wavenumber σ_0 and unit radiance, then a spectral line of infinitely small width $d\sigma$, radiance $L_\sigma(\sigma)\,d\sigma$, and wavenumber σ is recorded as a function $L_\sigma(\sigma)A(\sigma'-\sigma)\,d\sigma$. The whole spectrum emitted by the source therefore gives rise to a function

$$S(\sigma') = \int_0^\infty L_\sigma(\sigma)A(\sigma'-\sigma)\,d\sigma$$

The recorded function is thus the convolution of the source function and the instrumental profile of the spectrometer.† The relationship is conventionally written:

$$S = L_\sigma * A \qquad (6.1)$$

6.4. Factors influencing the instrumental profile

The form of the instrumental profile is determined by several factors peculiar to the instrument. In the case of a slit spectrometer, diffraction and the widths of the two slits are clearly the important factors; aberrations must nearly always be taken into account as well.

† The performance of dispersing instruments may be compared with imaging systems which reproduce a distribution of light intensity $O(x,y)$ (the object function) as a distribution of illumination in the image plane

$$I(x',y') = O(x,y) * D(x',y')$$

The spread function $D(x',y')$ which represents the degradation undergone by the optical information as it traverses an imperfect optical system is, in the special cases of spectroscopy, the instrumental profile.

The contribution of these factors may be specified by ascribing to each its own instrumental profile; this is the instrumental profile that would apply if all the parameters were negligible excepting the one under consideration. The inclusive instrumental profile $A(\sigma')$ is the result of all the causes of degradation acting simultaneously on the function $L_\sigma(\sigma)$ during transmission through the instrument; it is related to the individual functions $A_1, A_2, A_3, \ldots$ by the equation

$$A = A_1 * A_2 * A_3 * \cdots \tag{6.2}$$

that is, the instrumental profile is produced by convoluting the several individual functions. This follows from the properties of convoluted functions. Suppose that two broadening factors, represented by functions A_1 and A_2, operate successively on the function $L_\sigma(\sigma)$; the recorded output will be

$$S = (L_\sigma * A_1) * A_2$$

but as the convolution operation is, in the present case, associative and commutative, this may be written

$$S = L_\sigma * A_1 * A_2 = L_\sigma * A_2 * A_1$$

It follows that equation (6.2) is valid whatever the order in which the constituent functions $A_1, A_2, \ldots$ take effect.

6.5. *Definition of resolving power and luminosity of a spectrometer*

The definition of *resolving power* is identical with that used for spectrographs. As before, $\mathscr{R} = \sigma/\Delta\sigma$, $\Delta\sigma$ being the resolved spectral interval—that is, the minimum wavenumber interval between two radiations capable of being separately recorded by the instrument. We have already noted that there cannot be an absolute criterion for resolving power, since the ability to separate two neighbouring radiations depends on many factors, of which the most important is their relative intensities. If the simplest case is taken of two radiations having similar intensities and widths small compared with that of the instrumental profile then clearly the width of the instrumental profile will limit resolution.

The width of the instrumental profile has thus come to be accepted as the resolved spectral interval, so that the resolving power $\mathscr{R} = \sigma/\Delta\sigma$ is at once defined. This is, of course, the effective resolving power of the spectrometer, not to be confused with the intrinsic resolving power $\mathscr{R}_0$ of the dispersing element, which is always greater than $\mathscr{R}$.

The definition of *luminosity* adopted for spectrographs cannot be applied without modification to spectrometers because the photometric magnitude measured by the receiver is that of a different quantity. This quantity is now the luminous flux Φ, not the intensity of radiation or irradiance. If the instrument receives radiation in an infinitely narrow spectral waveband $d\sigma$, the transmitted flux is proportional to the monochromatic radiance $L_\sigma \, d\sigma$ of the source. The instrument will of course have been adjusted so that this flux is a maximum, the waveband $d\sigma$ being centred on the peak of the instrumental profile.

A logical definition of the luminosity is obviously given by the ratio of flux at the receiver to the source radiance in the band $d\sigma$:

$$\mathscr{L} = \frac{d\Phi}{L_\sigma \, d\sigma} \tag{6.3}$$

This ratio does in fact only depend on the characteristics of the spectrometer. Since the recorded function (6.1), given in the present case by $L_\sigma A(\sigma' - \sigma) \, d\sigma$, is proportional to the flux $d\Phi$, it follows that the luminosity is proportional to the peak height of the instrumental profile curve.

6.6. Instrumental profile of a slit spectrometer: choice of the relative width of the slits

When a stigmatic spectrometer receives a perfectly monochromatic radiation an image of the entrance slit is formed in the focal plane of the focusing lens and the purpose of the scanning mechanism is to traverse this image across the exit slit. The effect of diffraction on the distribution of illumination in the focal plane of the focusing lens must of course be taken into account; reference may be made to the curves given in Fig. 2.7, which reproduce this distribution for various entrance slit widths. The effects of aberrations need not be considered since, as will be shown later, they can always be made negligible.

The instrumental function of a slit spectrometer has, therefore, three components arising respectively from diffraction and the widths of the two slits. Since these may be considered in any order, the contribution from diffraction can be put aside for the moment and attention concentrated on the influence of the two slits. The function A_2 corresponding to the exit slit can be readily formulated: it is the function that would be recorded with an infinitely narrow entrance slit illuminated by a monochromatic source and with an optical system free from diffraction; this amounts to saying that the image of the entrance slit would be infinitely narrow. As this image is scanned across the exit slit (of which the angular width is α_2) it is obvious that the transmitted flux remains constant so long as the entrance slit image falls within the exit slit and that when the image is outside that region the flux is zero. Function A_2 is therefore a rectangular slit function. Its width is easily calculated. Let $d\theta_2/d\sigma$ be the angular dispersion; two incident radiations differing in wavenumber by $\Delta\sigma$ will give rise to two emergent rays with an angular separation of $(d\theta_2/d\sigma)\Delta\sigma$. So, when the emergent ray corresponding to a monochromatic radiation scans through an angle α_2 the setting wavenumber σ' varies by

$$\Delta\sigma'_2 = \frac{\alpha_2}{d\theta_2/d\sigma}$$

This last quantity represents the width of the function $A_2(\sigma')$. In the same way it is evident that the instrumental profile $A_1(\sigma')$ determined by the entrance slit is a rectangular function of width

$$\Delta\sigma'_1 = \frac{\alpha_1}{d\theta_1/d\sigma}$$

α_1 standing for the angular width of this slit and $d\theta_1/d\sigma$ the angular dispersion which would be applicable to rays passing from the exit to the entrance slit.

The angular dispersions $d\theta_1/d\sigma$ and $d\theta_2/d\sigma$ which correspond to the two directions of propagation through the instrument are not always equal; they are only so for a prism operating at minimum deviation or in an autocollimator mounting and for a grating used in such a mounting;† since, however, these cases are by far the most common we shall, from now on, assume equality of the two angular dispersions in order to simplify the discussion. This will in no way restrict its generality.

The rectangular functions A_1 and A_2 thus have widths proportional to α_1 and α_2 (Fig. 6.2); their heights are equal.

To determine the convolution $A_1 * A_2$ it is only necessary to move curve (1) laterally over curve (2) in small steps and to

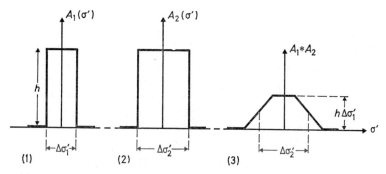

Fig. 6.2. Instrumental profile of a slit spectrometer, for the case where the angular widths of the two slits differ.

measure, at each step, the common area contained between the two curves and the σ'-axis. This amounts to saying that the flux transmitted by the spectrometer is proportional to the area outlined by the exit slit on the image of the entrance slit.

The convolution $A_1 * A_2$ is represented by a trapezium (Fig. 6.2) whose height is proportional to the angular width of the narrower slit (S_1 in the Figure) and whose width at half-height is proportional to the angular width of the broader slit. Under these conditions the function $A_1 * A_2$ will have, for a given width, a maximum height if $\Delta\sigma'_1 = \Delta\sigma'_2$, thus requiring the angular widths α_1 and α_2 of the slits to be equal. This condition should therefore always be satisfied, since it ensures the highest possible luminosity at a given resolving power. The profile of the function $A_1 * A_2$ is then triangular.

† Expressed more generally, the angular dispersions measured in the two directions of propagation are equal when the dispersing element does not act as an anamorphic component (see footnote, p. 13).

The effect of diffraction must now be considered. It may be represented by a function A_3 which expresses the distribution of illumination in the focal plane of the focusing lens when the entrance slit is infinitely narrow; it has the standard form expressed by equation (2.2). The inclusive instrumental profile A of the spectrometer is now the convolution of A_3 with the triangular function derived as above. The shape of A depends largely on the relative widths of $A_1 * A_2$ and A_3, hence on the angular widths of the slits compared with that of the diffraction spread, the latter being λ/a in the customary notation.

In particular, it is clear that if α is much greater than λ/a the effect of diffraction is negligible. The height and width of the instrumental profile are then proportional to α, as before. Since the luminosity $\mathscr{L}$ of a spectrometer is proportional to the height of its instrumental profile and its resolving power $\mathscr{R}$ inversely proportional to the width of the profile† it can now be stated that in the case of comparatively wide slits $\mathscr{L}$ varies directly, and $\mathscr{R}$ inversely, with the angular width of the slits.

When this angular width is no longer large compared with λ/a, the effect of diffraction must be taken into account. The instrumental profile is then intermediate between the triangle of the preceding case and the classical diffraction spread curve of an infinitely fine slit, tending towards the latter as α approaches zero.

6.7. *Variation of resolving power and luminosity with angular width of the slits*[9]

Now that we know how the instrumental profile changes with the angular width of the slits, it is easy to evaluate the influence of slit width on the resolving power and luminosity of a spectrometer: the curves of Fig. 6.3 indicate the form of the relationships.

6.8. *Comparison between spectrometers and spectrographs*

The curves of Fig. 6.3 may be compared with those of Fig. 2.8, which represent the variations of the same quantities in the case

---

† For convenience of comparison, profile widths are conventionally measured at an ordinate representing 0·405 of the maximum as in the case of diffraction.

of spectrographs. One difference is immediately obvious in the luminosity curves: while the luminosity of a spectrograph tends rapidly to the limiting value $\mathscr{L}_0$ when the slit width increases, here $\mathscr{L}$ increases indefinitely with α. Thus, luminosity can be increased without limit, though only at the expense of resolving power. Inversely, resolving power may be made to approach

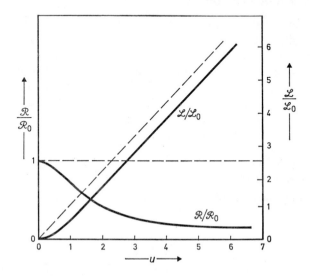

FIG. 6.3. Variations of resolving power $\mathscr{R}$ and luminosity $\mathscr{L}$ of a slit spectrometer with angular width of slits. The normalized variable u is the ratio of the angular slit width α to the angular width λ/a of the diffraction spread function for an infinitely narrow slit. $\mathscr{R}_0$ is the intrinsic resolving power of the dispersing element and $\mathscr{L}_0$ the luminosity with normalized slit widths $u = 1$, in the absence of diffraction.

the intrinsic resolving power $\mathscr{R}_0$ if luminosity is a secondary consideration—a second advantage gained over the spectrograph by virtue of the elimination of the photographic plate. The graininess of the emulsion often limits the resolving power of a spectrograph to a value well below $\mathscr{R}_0$ (the curves of Fig. 2.8 are then not applicable), while the limiting value $\mathscr{L}_0$ of luminosity is imposed by the aperture of the focusing lens, so that in practice the opportunities for varying the characteristics of a spectrograph are very limited.

104 SPECTROSCOPY AND ITS INSTRUMENTATION

The spectrometer may be classed as a universal instrument in the sense that both resolving power and luminosity may be varied over very wide limits, provided always that the reciprocal dependence is acceptable.

6.9. *Evaluation of luminosity*

We have so far considered only the relative variations of luminosity with angular slit width. The equation expressing $\mathscr{L}$ in terms of its parameters is easily derived. Ignoring diffraction for the moment and assuming the entrance slit to be illuminated by monochromatic radiation of intensity I and to be imaged onto the exit slit, the emergent flux Φ can be expressed in terms of I and the instrument parameters:

$$\Phi = \tau L S \Omega = \tau L U$$

where τ is the transmission factor of the instrument, S the usable area of the focusing lens O_2 and Ω the solid angle subtended by the exit slit at the image nodal point of the lens.

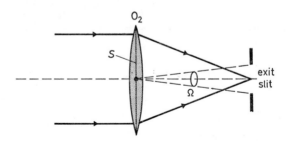

FIG. 6.4. Parameters of the beam at the exit slit.

Hence

$$\mathscr{L} = \Phi/L = \tau S \Omega = \tau S \alpha \beta \qquad (6.4)$$

β being the angular height of the slits. As might be expected, luminosity proves to be proportional to the angular width of the slits under conditions where diffraction is negligible.

It is important to realize that *the characteristics of the two lenses O_1 and O_2 do not appear in the expressions either of luminosity or of resolving power* (the dimension S is determined

SPECTROMETERS AND MONOCHROMATORS 105

by the dispersing element, not the lens). Resolving power being dependent on *angular* slit width, the choice of the focal length of the lenses remains open; the concept of a focal length linked to a limiting resolving power, arising from the influence of emulsion grain size, has no significance here. Similarly, luminosity depends only on the angular dimensions of the slits and on two parameters τ and S that are determined by the dispersing element, apart from a small loss in transmission due to reflection and absorption by the lenses; the situation is thus radically different from that pertaining to spectrographs, where luminosity is essentially determined by the relative aperture of the focusing lens. *It is totally unnecessary to provide a spectrometer with lenses of very high aperture*, and the correction of aberrations is thus greatly facilitated. (Smaller relative apertures are compensated by larger slit widths.) Their diameters and focal lengths can therefore be chosen to suit immediately practical considerations, with fairly large focal lengths to avoid the necessity of very narrow slits; rigidity and compactness are the only limiting considerations. For the infra-red region there is another requirement, that of concentrating the flux onto as small a receiver area as possible.†

6.10. *Relationship of resolving power and luminosity*

Equations formalizing the relationship between $\mathscr{R}$ and $\mathscr{L}$ can be derived.[11] The expression for luminosity has already been obtained in the form $\mathscr{L} = \tau S \alpha \beta$, valid when the angular slit widths are large enough to justify omission of the effects of diffraction. Similarly, the effective resolving power is $\mathscr{R} = \sigma/\Delta\sigma$, the resolved spectral interval $\Delta\sigma$ being equal to the width of the instrumental profile, that is, to $\alpha/(d\theta/d\sigma)$ under the above conditions. Replacing α in equation (6.4) by its expression in terms of $\mathscr{R}$ gives

$$\mathscr{L} = \frac{\tau S \beta \sigma \, d\theta/d\sigma}{\mathscr{R}} \qquad (6.5)$$

† In the infra-red, the source of noise, which affects all measurements, originates in the detector and increases with the area of the receiving surface (see § 6.24).

We know already that $\mathscr{L}\mathscr{R}$ is invariant with respect to slit widths. The evaluation of this invariant will enable useful comparisons of different types of spectrometer to be made; for this purpose it is convenient to introduce the intrinsic resolving power $\mathscr{R}_0$ of the dispersing element. This is given by $a(d\theta/d\lambda)$ or $a\sigma^2(d\theta/d\sigma)$. Hence

$$\mathscr{L} = \frac{\tau S \beta}{a\sigma} \cdot \frac{\mathscr{R}_0}{\mathscr{R}} \qquad (6.6)$$

from which, since $\mathscr{L}_0 = \tau S\beta/a\sigma$ (when $\alpha = \lambda/a$), we have

$$\mathscr{L}\mathscr{R} = \mathscr{L}_0 \mathscr{R}_0$$

If h represents the height of the beam at its emergence from the dispersing element, measured perpendicularly to the direction of dispersion, the area S may be written $S = ah$, from which

$$\mathscr{L} = \frac{\tau h \beta}{\sigma} \cdot \frac{\mathscr{R}_0}{\mathscr{R}} \qquad (6.7)$$

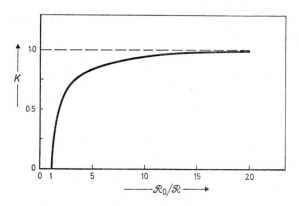

FIG. 6.5. Variation of the factor K by which expression (6.5) for luminosity must be multiplied to allow for effect of diffraction.

This expression for the luminosity of a slit spectrometer brings out the importance of the ratio $\mathscr{R}_0/\mathscr{R}$; when possible it always pays to use a dispersing element of which the intrinsic resolving power is well above the desired effective resolving power, because the luminosity is greater by the ratio $\mathscr{R}_0/\mathscr{R}$.

From this point of view, grating spectrometers are always to be preferred to prism instruments by virtue of their considerably higher resolving power.[11] This importance of the ratio $\mathscr{R}_0/\mathscr{R}$ is easily understood: for a given effective resolving power, the appropriate slit widths increase with the angular dispersion, and so with the intrinsic resolving power. The transmitted flux and, in consequence, the luminosity obviously increase with the slit widths.

The preceding results apply when the effect of diffraction is negligible; when the slit widths are too small for this to be true, luminosity is no longer proportional to $\mathscr{R}_0/\mathscr{R}$. The variation of $\mathscr{L}$ and of $\mathscr{R}_0/\mathscr{R}$ with angular slit width in the latter case is shown in Fig. 6.3. From the data given in that Figure are derived the values of a factor K (as a function of $\mathscr{R}_0/\mathscr{R}$) (Fig. 6.5) by which the expression (6.5) of luminosity must be multiplied to take account of the effect of diffraction.[11]

6.11. Comparison of prism and grating spectrometers[10, 11]

P. Jacquinot has established the relationship between the luminosity $\mathscr{L}_P$ of a prism spectrometer and that, $\mathscr{L}_G$, of a grating spectrometer of the same effective resolving power. If the transmission factor τ and the angular heights β of the slits are of the same order for the two instruments, then, from equation (6.7)

$$\frac{\mathscr{L}_P}{\mathscr{L}_G} = \frac{h_P}{h_G} \cdot \frac{(\mathscr{R}_0)_P}{(\mathscr{R}_0)_G}$$

the suffixes P and G characterizing quantities relating to prism and grating instruments, respectively.

The intrinsic resolving power of a grating is (§ 4.3, equation 4.10),

$$(\mathscr{R}_0)_G = l(\sin\theta_2 - \sin\theta_1)/\lambda$$

That of a prism is (§ 3.2)

$$(\mathscr{R}_0)_P = e\, dn/d\lambda$$

Hence

$$\frac{(\mathscr{L})_P}{(\mathscr{L})_G} = \frac{h_P e}{h_G l} \cdot \frac{\lambda \, dn/d\lambda}{(\sin\theta_2 - \sin\theta_1)}$$

$h_P e$ is the area of the prism base, $h_G l$ the ruled area of the grating.

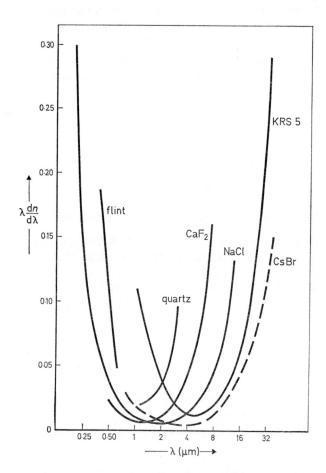

FIG. 6.6. Variation, as a function of wavelength, of the quantity $\lambda \, dn/d\lambda$, for various materials. This quantity represents the ratio between the luminosity of a spectrometer equipped with a prism of the particular material and that of a grating spectrometer of the same effective resolving power (after P. JACQUINOT, *J. Opt. Soc. Amer.* 1954, **44**, 761–5).

Since it is logical to compare dispersing elements of similar dimensions we may put $h_P e = h_G l$; it is also reasonable to assume the blaze angle to be in the region of 30°, so that $\sin \theta_2 - \sin \theta_1 \approx 1$. We then have

$$\frac{\mathscr{L}_P}{\mathscr{L}_G} = \lambda \frac{dn}{d\lambda}$$

a quantity depending only on wavelength and on the material of the prism. It will be noted that the ratio $\mathscr{L}_P/\mathscr{L}_G$ is practically always less than 0·10 and may even fall to much lower values. This means that a grating spectrometer has a luminosity at least ten times greater than that of a prism spectrometer of the same effective resolving power if the dispersing elements are of the same size in the two cases.

PRISM SPECTROMETERS

In spite of their distinctly poorer performance in terms of resolving power and luminosity, prisms have by no means been ousted by gratings in spectrometers. The spectral region in which they can be used covers part of the ultra-violet, the visible, and the infra-red as far as about 40 μm.

The conclusions of § 3.2 relating to the shape and dimensions of prisms remain valid for the case of spectrometers. Materials for the photographic range of wavelengths have already been discussed; it will be remembered that flint glass covers the visible range, quartz and fused silica the visible and ultra-violet down to 2000 Å. We now have to consider materials for an extension of the spectral range into the infra-red.

6.12. *Materials transparent in the infra-red*[2]

The transparency of the optical materials commonly used in the visible region extends a short way into the near infra-red, to about 2·5 μm for flint glass and to 3·5 μm for fused silica and quartz.

Some glassy materials have recently been developed for the infra-red region; the best known are calcium aluminate (transparent to 5 μm) and arsenic trisulphide (long-wave limit, 12 μm), though the latter has, unfortunately, rather a low melting point (210 °C).

Prisms of various crystalline compounds have been in common use for many years, as have crystalline lenses and windows; their range extends to intermediate infra-red wavelengths. Calcium fluoride (fluorite, CaF_2) is suitable up to 9 μm; sodium chloride, very widely used, has an upper limit at about 15 μm but suffers from the two disadvantages of being soft and hygroscopic, so it is not easy to polish and to maintain in good condition. Above 15 μm, the available materials are potassium bromide (25 μm), caesium bromide and caesium iodide (both reaching 40 μm) but in these last two crystals the disadvantages found in sodium chloride are present to an even greater degree; in particular, they are extremely hygroscopic. Considerable use has also been made of the mixed crystal thallium bromoiodide, known as KRS-5, of which the upper limit is also 40 μm; this material has an undesirably high refractive index which causes a serious loss of radiation by surface reflection.

The exploitation of these materials in spectrometry has been greatly facilitated by the availability of synthetic monocrystals of large size free from the many defects, such as twinning and inclusions, that are often present in natural crystals.

Table 6.1 summarizes the characteristics of these crystals.

TABLE 6.1. Characteristics of crystals for infra-red prisms

Compound	Long-wave limit (μm)	Notes
Fluorite, CaF_2	9	
NaCl	15	Soft, hygroscopic
KBr	25	Very hygroscopic
CsBr, CsI	40	Very hygroscopic
KRS-5	40	High index, reflection losses

Materials of an important new type have made their appearance in recent years; in their physical and mechanical properties they are a great improvement on the single crystals described above. They are particularly suitable for lenses and windows in the near and intermediate infra-red.

These new materials are sintered forms of the commonly used optical materials such as magnesium fluoride, zinc sulphide, and so on. They are much stronger than the monocrystalline alkali halides, are not hygroscopic, and the majority have melting points above 1000 °C; they are in fact, very good raw material for the fabrication of infra-red optical components. A range is already commercially available under the trade name IRTRAN; Table 16.2 gives their composition, transmission limit and refractive index.

TABLE 6.2. Characteristics of IRTRAN materials

Designation	Composition	Long-wave limit	Refractive index
IRTRAN 1	MgF_2	9 μm	1·34
IRTRAN 2	ZnS	14·5 μm	2·20
IRTRAN 3	CaF_2	11·5 μm	1·39
IRTRAN 4	ZnSe	21·8 μm	2·40
IRTRAN 5	MgO	9·5 μm	1·66
T 12	BaF_2/CaF_2	12·0 μm	1·42

The manufacture of these new materials is still under development, particularly with the object of producing specimens of increasingly larger dimensions.

Certain semiconductors must also be added to the list of infra-red materials, notably silicon and germanium in a highly purified form; apart from some absorption bands these too are highly transparent between 2 μm and 50 μm.[12]

Finally, it has recently been observed that the majority of crystals that are transparent in the near and intermediate infra-red but normally absorbing in the far infra-red become transparent in this region at very low temperatures, in general between 4 K and 80 K.[8] Since these materials retain their dispersive property at such temperatures, it is possible that by virtue of this effect prism spectrometers may eventually be able to operate at wavelengths above 40 μm.

6.13. Principal mountings for prism spectrometers

In operating a spectrometer it is convenient to keep the exit slit in a fixed position. The favoured mountings are therefore those giving constant deviation. One way of scanning the spectrum with such a mounting is simply to rotate the prism, but then minimum deviation occurs for only one wavelength setting. Minimum deviation is easily maintained over the whole wavelength range by including a plane mirror which turns with the prism (Wadsworth mounting). This mirror may, for example, be placed so that the incident ray that is to be refracted at minimum deviation undergoes a total deflection of 90° through the combined effects of refraction and reflection (Fig. 6.7). Once the components are set up for this condition at any one wavelength, it is easy to show that the same condition holds

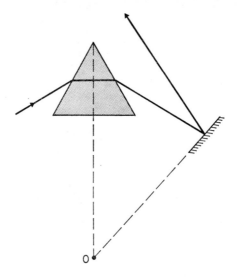

FIG. 6.7. Wadsworth mounting.

throughout the spectrum; if, for example, prism and mirror are turned through an angle α, the angle of incidence i at the prism is reduced by α; the minimum deviation $D = 2i - A$ therefore falls by 2α but, since the mirror has also turned through α in the same sense as that of the refracted ray, it is clear that the reflected ray does not change its direction. If the axis of rotation is suitably chosen, lateral displacement of the reflected ray may be prevented; the correct location is on the intersection of the mirror plane and the plane bisecting the prism.

This constant-deviation mounting is in general use. The prism

SPECTROMETERS AND MONOCHROMATORS 113

is sometimes cut so that one face acts as the mirror, as in the Pellin–Broca prism (Fig. 6.8) which is equivalent to a combination of two 30° prisms and a total-reflection prism. The deviation in this prism is a constant 90°; Fig. 6.8 illustrates its use for spectroscopes and small spectrometers designed for the visible region.

For the various reasons already explained, mirrors are more commonly used than lenses for the collimating and focusing

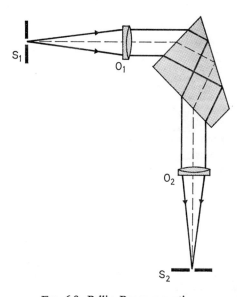

FIG. 6.8. Pellin–Broca mounting.

elements. A typical mirror-based mounting is shown in Fig. 6.9(a) (Czerny–Turner mounting) in which will be recognized the Z layout, already noted as a spectrograph system (Figs. 5.1 and 5.3). The virtue of this layout is that if the two spherical mirrors M_1 and M_2 are of equal focal length (as they would be in a spectrometer) and if the angles of incidence are equal, then coma induced by the first is annulled by equal and opposite coma introduced by the second. It must, however, be noted that coma compensation only occurs when (taking into account any reversals by auxiliary mirrors) the system is a true

Z mounting; in an arrangement such as that of Fig. 6.10(a), for example, the coma aberrations are additive, since it is equivalent to the basic form shown in Fig. 6.10(b) instead of the true Z form of Fig. 6.9(b).

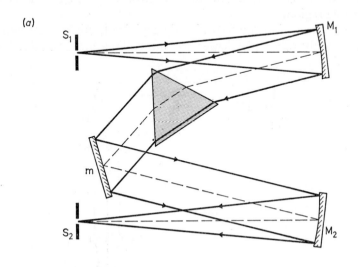

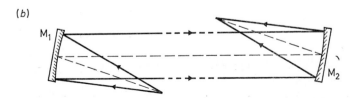

FIG. 6.9(a). Prism spectrometer: Czerny–Turner mounting.

FIG. 6.9(b). Z mounting, giving compensation of mirror aberrations.

The Littrow mounting is also fairly popular (Fig. 6.11). The spectrum is usually scanned by rotating a plane mirror m; in such an arrangement, the prism is only in use at minimum deviation for a single wavelength. Small auxiliary mirrors close to the slits to deflect the beams through 90° allow the concave mirror M to be used in the on-axis position, so avoiding off-axis aberrations and permitting the use of a spherical mirror. The only disadvantage is a partial obscuration of the dispersing prism.

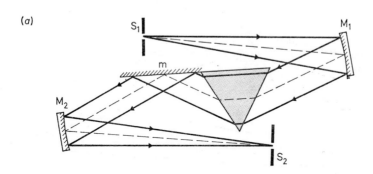

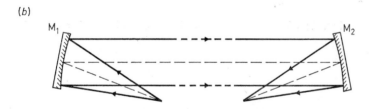

FIG. 6.10(a). Mounting to be avoided; aberrations uncompensated, as shown in Fig. 6.10(b).

FIG. 6.10(b). Mounting in which aberrations due to inclination of mirrors are additive.

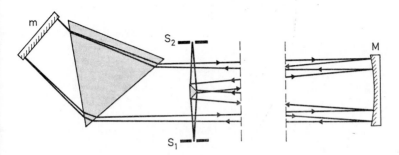

FIG. 6.11. Prism spectrometer. Littrow mounting.

An alternative to the preceding mounting is shown in Fig. 6.12. Here the mirror M is in an off-axis position and consequently, if the mirror is spherical, the off-axis aberrations are of the same sign; in this case it is necessary that the mirror should be

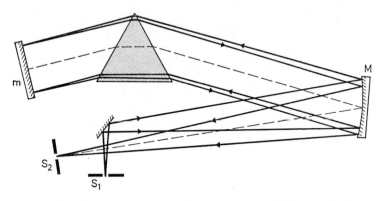

FIG. 6.12. Prism spectrometer. Littrow mounting. M is an off-axis paraboloid mirror.

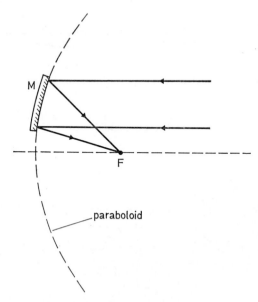

FIG. 6.13. Off-axis paraboloid mirror.

strictly stigmatic for an off-axis point source at infinity. An off-axis paraboloid is therefore called for, the reflecting surface being a section of a paraboloid of revolution not containing the vertex (Fig. 6.13).

GRATING SPECTROMETERS

Grating spectrometers are classified by the type of grating for which they are designed; plane grating instruments have advantages which give them almost a monopoly of the infrared, visible, near- and middle-range ultra-violet regions, whereas concave gratings must be used in the far ultra-violet, where no known mirror material has a sufficiently high reflectivity.

6.14. *Plane grating spectrometers*

To take full advantage of the blaze effect, plane gratings are always oriented so that the diffracted beam makes only a small

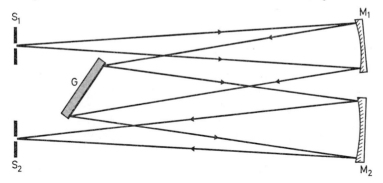

Fig. 6.14. Plane grating spectrometer. Czerny–Turner mounting.

angle with the (reversed) direction of the incident beam. Furthermore, the collimating and focusing elements have to be mirrors rather than lenses to avoid restricting the usable spectral range.

Under these conditions, the Czerny–Turner is again the favoured basic mounting (Fig. 6.14), having the advantages already described in connection with its use in prism spectrometers. The assembly comprising the two mirrors M_1 and M_2

is often replaced by a single mirror M, with a consequent gain in rigidity. This modification, first used by Fastie, is a direct derivative of the Ebert spectrograph (§ 5.3), as indicated by

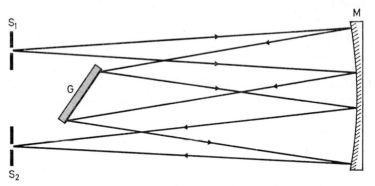

FIG. 6.15. Plane grating spectrometer. Ebert–Fastie mounting.

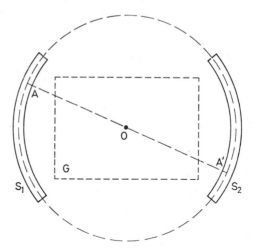

FIG. 6.16. Curved slits for the Ebert–Fastie spectrometer.

the name Ebert–Fastie given to the spectrometer.[3,4,5] The advantages of this mounting are even more marked in the constant-deviation spectrometer than in the spectrograph: since the slits S_1 and S_2 are always symmetrically disposed about the optical axis, the correction of coma is more nearly complete.

SPECTROMETERS AND MONOCHROMATORS

This correction would be perfect if the incident and diffracted beams were exactly symmetrical with respect to the mirror axis; this cannot be so because the two beams are inclined at different angles to the normal at the grating surface and consequently have different widths. Coma correction would, strictly, only be perfect if the grating were set for zero-order diffraction. This, of course, is of no practical use, but under normal working conditions coma in this type of mounting is slight. The only significant aberration is astigmatism, but this is not troublesome if the precaution is taken of using curved slits centred on the instrument axis and of properly chosen radius. Corresponding to a point source such as A (Fig. 6.16) there will be a line image A' perpendicular to the line AOA' and therefore tangential to the exit slit S_2. Under these conditions, the slits can have a considerable angular height β, so giving the spectrometer a high luminosity (equation 6.4). When, for some special reason, it is desirable to avoid astigmatism, a Littrow mounting analogous to that described above for prism spectrometers (Fig. 6.12), must be employed.

6.15. *Application of gratings with variable-depth grooves*

The luminous flux diffracted by an échelette grating is generally raised to a maximum by the blaze effect when both the incident and diffracted beams are normal to the facets (Fig. 4.7, § 4.5). It is, however, obvious that with the usual mountings this can only happen, in a given order, for a particular wavelength since the grating has to be rotated to scan the spectrum. The efficiency of the grating (ratio of diffracted to incident flux) is thus a maximum in the kth order for a particular wavelength λ_0 and decreases on either side of λ_0.

To maintain efficiency at its highest value over the whole spectrum, the rays have to remain normal to the facets. This is of course impossible with a normal grating but by the use of a grating having grooves of variable depth the condition can be met. The solution is based on a purely mechanical principle and is only applicable in the infra-red where the tolerances are wide enough to be met by precision engineering methods.

The variable-depth grating is made up of a pile of thin

platelets having accurately plane parallel faces, held together as shown in Fig. 6.17 by G and G'. Their upper edges AB, CD, ... act as the working facets of this grating.[21]

The orientation of the facets remains constant and the incident beam is always normal to the facets; the spectrum is scanned by rotating the flat supporting plate on which the platelets rest (its axis is at O, Fig. 6.17), so uniformly changing the height of each step along the grating surface.

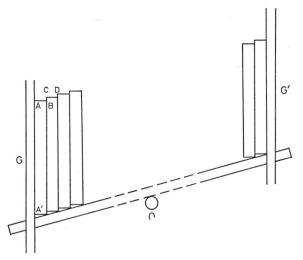

FIG. 6.17. Variable-depth grating.

This device appears to be very elegant in principle, but when account is taken of the precision known to be necessary in ruling a grating (see § 4.8, p. 57) it will be realized that the construction of variable-depth gratings must be extremely difficult. It seems unlikely to be practicable in any wavelength region below that of the far infra-red.

6.16. *Separation of spectra of differing orders*

The problem of separating spectra of differing orders has to be solved for spectrometers just as for spectrographs: methods similar to those already described (see Chapter 5) are applicable. In the present case it is naturally no longer possible to register several orders simultaneously since the spectrum now has to be scanned over a finite period of time.

Systems for separating orders have particular importance in the far infra-red because in that region the free spectral range $\Delta\sigma_0$ for a grating is always comparatively small as a result of the large pitch of the rulings. Thus a grating designed for a region in the neighbourhood of 500 μm has a pitch of about 1 mm and, with a blaze angle of 30°, this gives a value of only 10 cm^{-1} for $\Delta\sigma_0$. Thus, the spectral interval between $\lambda = 1$ μm and $\lambda = 1$ mm contains that free spectral range 10 000 times. This emphasizes the need for effective order separators, especially as sources of far infra-red radiation emit a high proportion of their energy at shorter wavelengths. An example given by A. Hadni[7] for the case of an échelette grating having 14 grooves per mm shows that the higher diffracted orders deliver to the exit slit 20 000 times the energy carried by a first-order spectrum at $\lambda = 43$ μm.

To eliminate the parasitic radiation, auxiliary gratings are often used; these either transmit or reflect the desired radiation band in the zero order. The gratings are échelette gratings or, in the very far infra-red, grilles made of parallel metallic wires. The échelettes are generally formed in plastic materials such as polythene and are used in transmission. The efficiency curve of these gratings for zero order indicates that they are particularly useful for order separation, their reflection or transmission factor being close to unity for all wavelengths above a critical value λ_c and almost zero when λ is below λ_c.

Other types of filter are also employed, particularly plates of crystalline materials having suitably chosen transmission or reflection properties. These plates are sometimes used in the construction of selective modulators for the longer wavelengths. The modulator is in the form of a disk having open sectors alternating with sectors of a material transparent in the near infra-red but opaque in the far infra-red. The transmitted flux is thus modulated only at the longer wavelengths and the corresponding photoelectric output is easily separated from the unwanted d.c. component by a tuned amplifier. Diffusing metal and silica surfaces are also effective in selecting the longer wavelengths if the dimensions of the microstructure of the diffusing surface are small enough to allow the longer wavelengths to be specularly reflected. Christiansen filters are also very effective; for the present purpose they are usually composed of microcrystals embedded in a plastic pellicle. If these techniques are judiciously combined, the parasitic radiations in any one spectral region can be quite effectively eliminated.

One other method of separating spectra of different orders remains to be described; it is based on a principle quite different from those described so far. This is the method of modulation by interference proposed by J. Strong,[24] in which the diffracted radiations of different orders are modulated at different frequencies so that the corresponding detector outputs are separable by electronic filters. As we shall find later, a two-beam interferometer can act as a modulator, since the emergent flux is of the form

$$\Phi = \Phi_0\{1+\cos(2\pi\sigma\delta)\} = \Phi_0\{1+\cos(2\pi\sigma vt)\}$$

if the path difference δ varies with the velocity v. The frequency of modulation σv is thus proportional to the wavenumber σ of the radiation. In particular, Strong suggested as a modulator a variable-depth grating which, in its zero order, is equivalent to a two-beam interferometer.

6.17. *Concave grating spectrometers*

As in the case of spectrographs, spectrometers designed to operate in the far ultra-violet beyond 1000 Å must use concave gratings. No material has a reflectivity high enough in this region to make the use of mirrors possible.

One of the major problems in designing a concave grating spectrometer arises from the requirement for accurate focusing over a wide spectral range without adjustment of the position of the entrance and exit slits.

The simplest solution is to move the grating along the Rowland circle (Fig. 6.18).[24] The position of the circle thus remains constant as the spectrum is explored and so the slits S_1 and S_2 may also remain in fixed positions, the focusing condition being always satisfied. Two serious disadvantages of this system are that the incident beam moves across the grating and the emergent beam changes direction. The latter effect can be eliminated by placing a fixed diaphragm in front of the grating; to be effective its aperture must be small compared with the area of the grating, but this results in poor utilization of the grating.

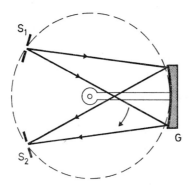

FIG. 6.18. Concave grating spectrometer with grating rotatable about centre of Rowland circle (radius mounting).

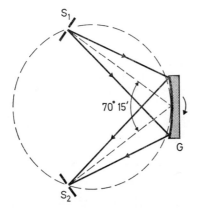

FIG. 6.19. Concave grating spectrometer. Seya–Namioka mounting.

A different solution was suggested by Seya[23] and put into practice by Namioka.[14] In their mounting the slits S_1 and S_2 are both fixed and the grating G (Fig. 6.19) turns about an axis parallel to the rulings and passing through its apex. As a result of this rotation, the Rowland circle moves away from the slits, but Seya showed that when the angle between the incident and emergent rays is 70° 15', the error in focusing as the spectrum is scanned is so small that it remains negligible for a scan covering the whole ultra-violet range.

The Seya–Namioka mounting offers the advantage of a very simple mechanical design; moreover, the wide separation of the slits facilitates the attachment of sources, detectors and auxiliary equipment. There is unfortunately a corresponding disadvantage in that the high angles of incidence at the grating introduces considerable astigmatism.

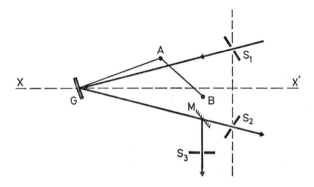

FIG. 6.20. Concave grating spectrometer. Pouey–Romand–Martin mounting (C.N.R.S. patent). Mirror M is removable, providing for alternative exit slits S_2 and S_3.

This astigmatism can be reduced by reducing the angle of incidence to quite a low value, as in the Eagle mounting (see § 5.10). It is then necessary to introduce a second movement of the grating to maintain proper focusing at the exit slit as the spectrum is scanned: as the grating is rotated it must be moved along the bisector of the angle between the incident and emergent rays. Various mechanisms have been proposed to coordinate these two motions.[16,17,18] Fig. 6.20 is a diagram of one such mechanism patented by Pouey, Romand and Martin (C.N.R.S. patent).[16] The linking of the rotation of the grating about an axis passing through its vertex (perpendicular to the plane of the figure) and its translation along the line XX' is provided by the two arms GA and AB, joined by a pivot at A.

The grating is mounted on GA so that GA is normal to its surface while AB is also pivoted at a fixed point B. A thorough geometrical analysis has established the lengths of the arms GA and AB and the position of the fixed point B that together give optimum focusing over the widest possible spectral range. The short-wave end of the range is about 400 Å; below this the reflectivity of the grating falls to too low a value.

To penetrate the region below this short-wave limit, the grating must be used at grazing incidence, as in the case of spectrographs (see § 5.11). Here again the movements of the grating have to be controlled by suitably designed mechanisms.[1, 19, 22]

GRILLE SPECTROMETERS

6.18. *Introduction*

All the prism and grating spectrometers so far described have entrance and exit slits to isolate a small spectral interval within the spectrum produced by the dispersing element. These slits carry with them the disadvantage of very severely limiting the étendue of the beam accepted by the instrument and consequently of restricting the flux reaching the detector. As a result, the angular slit width and the luminosity vary inversely with the resolving power.

The grille spectrometer was devised by A. Girard with the object of avoiding the use of slits and so making the étendue of the beam, and hence the luminosity, independent of the resolving power.[6]

6.19. *Principle of the grille spectrometer*

Consider the distribution of radiance in the front focal plane of the collimator of a spectrometer having an entrance slit S_1 of width l. Along an axis $0x$ perpendicular to S_1, the variation of radiance is represented by a function $S(x)$ which is constant over a length equal to l and zero outside this region. $S(x)$ is the familiar slit function and its Fourier transform is $G(u) = l\{\sin(\pi u l)\}/\pi u l$ (see § 4.2). The curve representing $G(u)$ (Fig. 4.4) is composed of a central portion, of which the width measured along the u-axis is $2/l$, and of secondary maxima of rather rapidly decreasing amplitudes.

The slit function $S(x)$ may therefore be considered as being equal to the sum of an infinity of sinusoidal functions, all in phase at $u = 0$, of which the spatial frequencies are spread over the range $u = 0$ to $u = \infty$.† Although the frequency band is infinitely wide the form of $G(u)$ makes it reasonable for practical purposes to limit the bandwidth to $1/l$; in other words a slit of width l can be regarded as having a pass band, in terms of spatial frequency, of about $1/l$.

This analysis leads to the concept of replacing the distribution $S(x)$ by a superposition of an infinity of sinusoidal distributions of radiance having spatial frequencies ranging continuously from zero to $1/l$. These distributions of radiance can be produced by uniformly illuminated filters of which the transmission varies sinusoidally along the x-axis. Theoretically, it would be possible to obtain a spectrum in the following way. Slit S_1 would first be replaced by a sinusoidal filter of period p ($p > l$) and slit S_2 by a second filter identical with the first; the spectrum would then be scanned by rotating the prism or grating and the detector output recorded in the usual way. This operation would then be repeated with a succession of filters having spatial periods evenly distributed between l and infinity. Finally the ordinates of all the recorded curves would be added. In this way, the spectrum obtained with slits of width l would be reconstructed but the luminosity ascribable to the hypothetical process would be much higher because the transparent area of the filters would be much greater than that of the slits.

Such a procedure is, of course, too tedious to be practicable, but it can be simplified by carrying out all the recordings simultaneously; to do this a series of filters of different periods can be juxtaposed in the front focal plane of the collimator.

Assuming bars of these filters to be vertical (that is, parallel to the y-axis, the usual direction of a slit), the available height H may be regarded as being divided into n equal segments, each occupied by a filter of a particular period (Fig. 6.21).[6]

† $G(u)$ being an even function, as is $S(x)$, it is not necessary to consider negative values of u.

The transmission factor of any one of the filters varies in the x-direction according to the law

$$\tau = \tfrac{1}{2}\{1 + \sin(2\pi x/p)\}$$

If now the number of filters tends to infinity, and if the spatial frequency $1/p$ is allowed to increase linearly with y, we have

$$\tau = \tfrac{1}{2}\{1 + \sin(2\pi kxy)\}$$

We have now arrived at the concept of a *grille* in which the curves of equal transmission are equilateral hyperbolae having the axes $0x$ and $0y$ as asymptotes. It would, however, be very

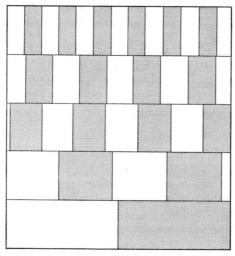

FIG. 6.21. Array of filters of different periods (after GIRARD[6]).

difficult to prepare screens having a sinusoidal variation of transmission in both x- and y-directions, as required by the equation above. Fortunately it is permissible to replace the sinusoidal variation by a square-wave variation—that is, to prepare a screen with alternately opaque and completely transparent zones bounded by equilateral hyperbolae as in Plate 6. The principle of the grille is not invalidated by this simplification.

If l is the smallest period present, the grille is equivalent to an infinity of filters having spatial frequencies ranging continuously from 0 to $1/l$ and so is almost equivalent, as far as pass band is concerned, to a slit of width l. The effective resolving power of a spectrometer fitted with grilles should therefore depend only on the minimum value of their period. The flux transmitted by a grille is clearly much greater than that passed by the equivalent slit and depends only on its area. The luminosity of a grille spectrometer must therefore be considerably higher than that of the corresponding slit spectrometer and independent of the resolving power. This is indeed the *raison d'être* of the grille spectrometer.

6.20. *Instrumental profile of the grille spectrometer*

Suppose the two slits of a typical spectrometer to be replaced by two identical grilles G_1 and G_2 of the form described in the previous section.

At a given instrument setting there will be a monochromatic radiation of wavenumber σ' for which the image G'_1 of G_1 is exactly superimposed on G_2. Following the terminology previously employed, σ' is the setting wavenumber. For this particular radiation, the flux passed by the entrance grille G_1 is totally transmitted by the exit grille G_2. For radiation of any other wavenumber, transmission is only partial.

At a point having coordinates x_2, y_2 in the plane of the exit grille the latter has a transmission factor $\tau(x_2, y_2)$. Now, an image G'_1 of the grille G_1 is formed in this plane but is displaced, relative to G_2, in the x-direction by an amount

$$x' = \frac{dx}{d\sigma}(\sigma - \sigma')$$

$dx/d\sigma$ being the linear dispersion measured in the plane of G_2. For radiation of wavenumber σ, the flux transmitted by an elemental area $dx_2\, dy_2$ of the exit slit is thus proportional to

$$\tau(x_2, y_2)\tau(x_2 - x', y_2)\, dx_2\, dy_2$$

The instrumental profile of the spectrometer is thus

$$A(\sigma - \sigma') = \frac{\iint \tau(x_2, y_2)\tau(x_2 - x', y_2)\, dx_2\, dy_2}{\iint [\tau(x_2, y_2)]^2\, dx_2\, dy_2}$$

the denominator being a normalizing factor.

For the grilles under discussion, the calculated values give a curve for $A(\sigma)$ of the form shown in Fig. 6.22, curve 1.

$A(\sigma)$ is essentially the sum of two functions:

(i) a very flat triangular function with a wide base on the σ-axis of length $2L/(dx/d\sigma)$, L being the width of the grille in the direction of dispersion;

(ii) a second function, also triangular but much narrower than the first. Its half-peak width is approximately $l/(dx/d\sigma)$, where l is, as before, the smallest period of the grille.†

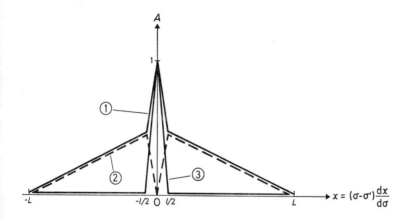

FIG. 6.22. Instrumental profile of a grille spectrometer. Alternation system.

An instrumental profile of such a great width is of course not desirable but there are two ways of modifying the instrument in such a way as to eliminate the wings of the profile.

The first method involves the successive use of two complementary exit grilles. By this is meant a pair of grilles G_2 and G'_2 of which G_2 has a transmission

$$\tau = \tfrac{1}{2}\{1 + \sin(2\pi kxy)\}$$

† The exact half-width depends on the contour of the grilles: this is chosen to give a suitable apodization (see § 7.31, p. 194) of the instrumental profile, but the order of magnitude is unaltered.

and G_2' a transmission

$$\tau' = \tfrac{1}{2}\{1 - \sin(2\pi kxy)\}$$

Physically this means that the opaque tracks on G_2' correspond to the transparent areas on G_2 and vice versa.

When grille G_2 is replaced by G_2' the instrumental function becomes that represented by curve 2 of Fig. 6.22; to eliminate the wings of the original instrumental function (curve 1) it is therefore only necessary to subtract the ordinates of curve 2 from those of curve 1, leaving curve 3. The instrumental function so obtained is a triangle of half-peak width $l/2(dx/d\sigma)$ and is therefore nearly the same as that of a spectrometer with two equal slits of width $l/2$.†

6.21. *Alternation system*

To put the method described into practice, we must derive a signal proportional to the difference of the values of the emergent flux with, first G_2, then the complementary grille G_2', in place. An elegant way of extracting these two signals is to make the opaque areas of G_2 reflecting: the reflected flux is that which would be transmitted by the complementary grille.

To register the difference signal it would be possible to record the transmitted and reflected beam intensities by means of two separate photodetectors; the disadvantage of this scheme is that the difference of two nearly equal fluxes is very sensitive to small variations in detector sensitivity. It is therefore preferable to use a method requiring only one detector which receives the transmitted and reflected beams in succession. The beam-switching is provided by a disk having equal transparent and reflecting areas in alternate sectors. The signal is thus modulated in square-wave form at the frequency of alternation and its amplitude is proportional to the difference of the two fluxes. A tuned amplifier followed by a rectifier eliminates the d.c. component and gives a measure of the alternating component.

In the infra-red the beam-switching disk would introduce a parasitic signal due to the thermal radiation emitted by the disk,

† These results would strictly be true only if diffraction did not occur.

this radiation being modulated at the same frequency as that of the measuring signal. The disk must therefore be placed in front of the spectrometer, operating in conjunction with the entrance slit instead of the exit slit; the principle is the same as for the previous system. The layout is then as shown in Fig. 6.23, in which the entrance grille G_1 is now operating in both transmission and reflection. A Littrow mounting must be used with grilles because a complete absence of astigmatism is essential for their proper functioning; this is not achieved with the Ebert–Fastie mounting. The collimating mirror M is parabolic.

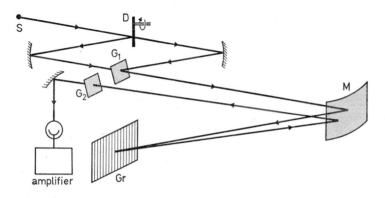

FIG. 6.23. Grille spectrometer. S: source. D: beam-switching disk. Gr: grating. G_1, G_2: grilles (after A. GIRARD, *J. Phys. Radium*, 1963, **24**, 141).

6.22. Oscillation system

In the preceding mounting, the difference between the intensities of two beams is measured. If this difference is to be an accurate measure of the monochromatic intensity of the source at the setting wavelength the two beams must be perfectly symmetrical, and this demands very precise location of the components.

In order to avoid this dependence on very accurate construction, Girard proposed a second system, the *oscillation* system.

Suppose the image of the entrance grille formed on the exit grille to be displaced by a given distance in the *y*-direction with

132 SPECTROSCOPY AND ITS INSTRUMENTATION

respect to the latter. The horizontal asymptotes of G_1' and G_2 will then be separated by an amount which can be taken to be at least equal to the smallest period l of the grilles. Calculation and experiment show that the instrumental profile is then reduced to curve 1 of Fig. 6.24 whereas when the asymptotes are coincident the profile is as curve 2, which is identical with curve 1 of Fig. 6.22. Suppose now that while recording the instrumental function the image G_1' is caused to vibrate in a direction parallel to $0y$; this movement is easily obtained by giving the collimating mirror a small oscillatory motion about its horizontal axis.

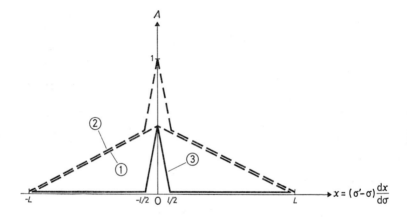

FIG. 6.24. Instrumental profile of a grille spectrometer. Oscillation system.

To determine the resulting instrumental profile, the entrance grille is illuminated with a monochromatic radiation of wavenumber σ and the setting wavenumber σ' is varied. While σ' remains considerably different from σ, that is, when the corresponding abscissa x in Fig. 6.24 is such that $|x| > l/2$, the emergent flux remains constant during a full oscillation period. When, however, σ' is close to σ ($|x| < l/2$) the emergent flux varies during a period of oscillation. The amplitude of modulation is equal to the difference of the ordinates of curves 2 and 1 in Fig. 6.24 and is shown as curve 3. If the photodetector output is amplified by an amplifier tuned to the modulation frequency and rectified, curve 3 is reproduced by the output signal and is, by definition, the instrumental profile.

The luminosity of a grille spectrometer operated on the oscillation system is only half that of one using the alternation system but its design is simpler and its adjustment less critical. Furthermore only one grille is required, used in reflection as the entrance grille and in transmission as the exit grille, or vice versa.

A point to be noted is that there is a fundamental difference in the modes of operation of a grille and a slit spectrometer. In the latter the exit slit isolates a narrow pass band and only the radiation within this band reaches the photodetector. In the grille spectrometer, on the contrary, the selected waveband is distinguished from the rest of the radiation passing through the spectrometer by the fact that only the flux within that band is modulated. The modulation amplitude is maximal at the centre of the pass band and serves to measure the spectral intensity of the source at the corresponding wavelength.

Another instrument using an analogous principle will be described in the next chapter: this is P. Connes' interference spectrometer based on wavelength selection by modulation amplitude.†

6.23. Étendue advantage

Replacing the slits of prism and grating spectrometers by grilles has achieved the aim of increasing, by a considerable amount, the étendue of the beam accepted by these instruments and of rendering it independent of resolving power.

The half-peak width of the instrumental profile is of the order of $l/2(dx/d\sigma)$; the effective resolving power is thus the same as that which would result from slits of width $l/2$, a width equal to the smallest transparent interval of the gratings. Note also that grilles make it possible to attain an effective resolving power equal to the intrinsic resolving power of the dispersing element by employing grilles of which the smallest spatial period is

† *French*: Spectromètre Interférentiel à Sélection par l'Amplitude de Modulation (S.I.S.A.M.).

equal to the width of the spread function. An equal result cannot be obtained with slits because these would have to have zero width.

The transparent area of a square grille of side L is $L^2/2$ while that of a slit of width l and height L is lL. The gain in étendue is therefore approximately $L/2l$; this is always a large factor and, of course, it increases as the resolving power is increased. Common values are $L = 30$ mm, $l = 0\cdot 1$ mm, giving a gain of 150. Experimental results, especially in the near and intermediate infra-red, confirm these theoretical gains.

6.24. *Influence of the type of detector*

The ability of a grille spectrometer to measure very weak radiation is not wholly characterized by the étendue of the accepted beam as it is in the case of slit spectrometers. This is due to the fact that the photodetector not only receives the very small flux within the resolved spectral region but is also flooded with unmodulated radiation from the neighbouring regions of the spectrum. Now, in certain cases this parasitic radiation can increase the background noise at the detector output, with the undesirable result that the detectability of weak radiations is reduced.

The noise at the output of the detector may have either of two origins according to the type of detector.

In the case of photoconductive detectors and thermal receivers, the noise is inherent in the detector itself; it varies, to a good approximation, with the square root of the area of the receiving surface and is practically independent of the incident radiation. With photo-emissive cathodes (as in photomultipliers), on the contrary, the noise due to the detector can be made negligibly small, to the extent that the photon noise due to the fundamental fluctuations in the received radiation can be observed.

It follows that in the infra-red the parasitic flux received by the thermal detector of a grille spectrometer does not, theoretically, increase the inherent noise level. If the étendue is increased, the signal, and hence the signal/noise ratio, should be increased by the same factor. In practice, the increase in étendue generally calls for some increase in the area of the receiving surface so that the real gain on replacing a slit with a grille is somewhat less than the theoretical value based on relative étendues but it is still considerable. In the visible and ultra-violet, where detectors are of the photo-emissive type, the unmodulated parasitic flux contributes so much photon noise that the advantage attaching to the grille is seriously diminished.

It is therefore only in the infra-red region that the grille spectrometer holds an undisputed advantage over the slit spectrometer.

MEASUREMENT OF ABSORPTION SPECTRA

6.25. *Introduction*

The instrumental requirements for absorption spectroscopy are basically similar to those found in emission spectroscopy and the same equipment is used in both cases. There is nevertheless a difference in the two techniques, arising from the fact that the flux absorbed by the substance cannot be measured directly; its value can only be deduced by comparing the incident flux Φ_i with the flux Φ_T transmitted by the sample; the absorption factor A is given by †

$$\Phi_T = (1-A)\Phi_i$$

from which

$$A = 1 - (\Phi_T/\Phi_i)$$

Charting the absorption spectrum therefore calls for two measurements of flux at each wavelength, one for the flux transmitted by the specimen and one without the specimen. The two measurements can be made point by point at selected wavelengths or by recording the continuous curves representing

---

† This simple equation assumes that reflection losses at the entrance and exit faces of the specimen are negligible; if, in practice, they are not negligible they are either evaluated separately or compensated experimentally.

$\Phi_i(\lambda)$ and $\Phi_T(\lambda)$ and then calculating A from the values at a number of points on these curves. This second method is naturally the faster but it is less reliable than the first. It assumes that the intensity of the source and the sensitivity of the detector remain constant during the recording of both curves; it also demands a high degree of reproducibility in the wavelength setting of the spectrometer. With the point-by-point method this latter quality is not important, while the source and detector are only required to be stable over a very much shorter time.

6.26. *Double-beam spectrophotometers*

The two preceding methods have one serious disadvantage: neither displays a graph of the absorption spectrum directly; the whole process of data logging and computation can be a lengthy one. To obtain a direct recording of the spectrum a double-beam spectrometer must be used. Fig. 6.25 shows a typical optical system.

The radiation from a source L is divided into two beams of equal angular width by means of mirrors m_1 and m_2; the first beam passes through the sample and the second serves as a reference beam. When the substance under examination is a gas or liquid contained in a cell C_1, an empty cell C_2 of identical construction is placed in the reference beam to compensate for reflection and absorption losses. A beam-switching disk D similar to that used in the grille spectrometer (see § 6.21) passes each beam in turn to the exit slit S; the square-wave signal generated by the alternated beams is amplified and rectified and results in an output proportional to the absorption factor A, since $A = (\Phi_2 - \Phi_1)/\Phi_2$. The method is commonly refined by automatically adjusting the reference flux Φ_2 to give a zero resultant, so taking advantage of the merits of a null method of measurement.

In the null method the reference and sampling beams may be balanced by introducing a graded attenuator At, calibrated in units of absorption factor, into the reference beam. Under these conditions the accuracy of measurement is unaffected by variations in the sensitivity of the photodetector or in intensity of the source. The only difficulty lies in the preparation of an attenuator of which the absorption function is reproducible and independent of wavelength. This is often ensured by employing a mechanical device such as a diaphragm of variable aperture shaped to give a suitable relationship between adjusting movement and attenuation; alternatively, a photographic wedge filter is sometimes used but this must have a perfectly neutral transmission characteristic. With such a system it is easy to arrange for automatic and continuous recording of absorption spectra: the amplified a.c. signal from the photodetector controls a servomotor which drives the attenuator

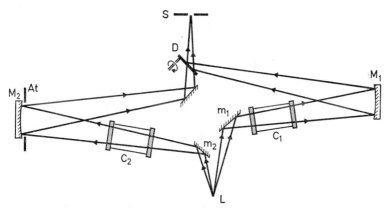

FIG. 6.25. Double-beam system for absorption spectrophotometer.

mechanism towards the balancing position. If the attenuator is also coupled mechanically to the recorder pen, the variation of absorption factor is plotted as the spectrum is scanned.

The two-beam system as shown in Fig. 6.25 is placed between the source L and the entrance slit S of the spectrometer. It could also be put between the exit slit and the detector but while this arrangement might be feasible for the ultra-violet or visible regions it is not practicable for the infra-red because the thermal radiation from the alternator disk D, itself modulated at the alternator frequency, represents a parasitic signal which will also be amplified. In the arrangement shown in Fig. 6.25 this parasitic signal is filtered out by the spectrometer.

SLIT MONOCHROMATORS

A monochromator is designed to extract from heterochromatic radiation a beam carrying radiation confined within a narrow spectral band of which the width and mean wavelength are controllable. Instruments for this purpose use prisms or gratings and indeed are very similar in construction to the corresponding spectrometers. Nevertheless, because of their different function, their characteristics must be examined from a different point of view.

6.27. *Simple monochromators*

We have seen that a spectrometer is essentially characterized by two quantities, resolving power and luminosity, both of which are directly influenced by the instrumental profile. The situation is similar for monochromators in that their function is to act as a filter, so that both the width of the pass band and the flux within that band are the important features. The term *monochromator* is in fact not a true description of the instrument, since when set for a particular wavenumber σ_0 the emergent flux is not strictly monochromatic but covers a narrow region of the spectrum centred on σ_0.

Just as the radiation entering a monochromator is characterized by the variation of its monochromatic radiance L_σ, the emergent beam is characterized by the variation, as a function of wavenumber σ, of the monochromatic flux transmitted† (or flux per unit wavenumber).

The emergent monochromatic flux is expressed by

$$\Phi_\sigma = L_\sigma \tau U$$

τ being the transmission factor of the instrument and U the étendue of the transmitted beam.

The monochromatic radiance of the source may be taken to be constant within at least a small spectral range. The shape of

---

† The definition of monochromatic flux Φ_σ is analogous to that of monochromatic radiance—that is, the flux carried by a beam within the spectral range $d\sigma$ is $d\Phi = \Phi_\sigma\, d\sigma$.

the curve giving the variation of Φ_σ as a function of σ is therefore determined by the variation of étendue U. If the monochromator is set at a wavenumber σ_0 we may write

$$\Phi_\sigma(\sigma - \sigma_0) = L_\sigma \tau U(\sigma - \sigma_0)$$

the factor $L_\sigma \tau$ being practically constant. The result is thus analogous with that obtained for spectrometers—that, for each value of $\sigma - \sigma_0$, Φ_0 is proportional to the area outlined on the monochromatic image of the entrance slit by the exit slit.

In spite of the resemblance of the two cases, the interpretation of this result is not the same for the monochromator as for the

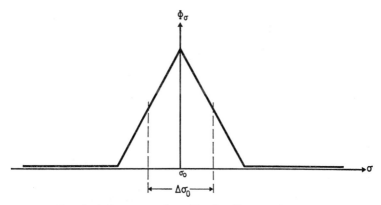

FIG. 6.26. Instrumental profile of a slit monochromator.

spectrometer. In the latter case the instrumental profile was plotted by rotating the dispersing element to scan the wavenumber region in the neighbourhood of a quasi-monochromatic spectral line. In the monochromator, however, the incident illumination has a continuous, or white, energy distribution and the dispersing element is stationary: the emergent radiation is described by a curve representing the variation of *monochromatic flux* as a function of wavenumber.

This variation is shown by, again, a curve in the form of a trapezium, identical with that of Fig. 6.2 when the angular widths of the two slits are different; when these widths are equal the curve is reduced to a triangle (Fig. 6.26) and it is clear that this setting is again the most efficient. The total flux transmitted by the monochromator is in fact proportional to the area between the curve and the σ-axis and reaches a maximum for a given half-width (that is, for a fixed pass band) when the trapezium is reduced to a triangle.

The principle for setting up a simple monochromator is thus the same as for a spectrometer: the angular slit widths should be equal for maximum transmission of flux at a given bandwidth.

6.28. Bandwidth and luminosity

With equal angular slit widths, the half-width of the pass band (or spectral range of the emergent radiation) is $\Delta\sigma = \alpha/(d\theta/d\sigma)$, α being the angular slit width and $d\theta/d\sigma$ the angular dispersion due to the dispersing element. The total emergent flux is $\Phi = \Phi_\sigma(\sigma_0)\,\Delta\sigma$, in which $\Phi_\sigma(\sigma_0)$ is by definition the maximum emergent monochromatic flux since it occurs at the centre σ_0 of the pass band. We also have

$$\Phi = L_\sigma\,\Delta\sigma\,\tau U_0 = L_\sigma\,\Delta\sigma\,\tau S\alpha\beta$$

where U_0 represents the maximum value of étendue permitted by the monochromator (for $\sigma = \sigma_0$); the quantities S and β are, as before (§ 6.9), the usable area of the exit lens and the angular height of the slits. Replacing α by its expression as a function of $\Delta\sigma$ gives

$$\Phi = L_\sigma\,\tau S\beta\,\frac{d\theta}{d\sigma}\,(\Delta\sigma)^2$$

In the present case where the source has a continuous spectrum, this indicates that the emergent flux varies as the square of the bandwidth.†

---

† It can be shown that for a line source of radiance L, the value pf Φ is

$$\Phi = L\tau S\beta\,\frac{d\theta}{d\sigma}\,\Delta\sigma$$

that is, the flux is proportional to the width $\Delta\sigma$ of the pass band of the monochromator.

SPECTROMETERS AND MONOCHROMATORS 141

It is particularly important to note that for a given bandwidth the flux depends only on the characteristics of the dispersing element and on the angular height of the slits. Contrary to what might be expected, it is in particular *independent of the relative aperture* of the instrument. This means that, as in the case of spectrometers, there is no point in employing high-aperture lenses, with the attendant problem of aberration correction.

Finally, since the flux is proportional to the angular dispersion, it is obvious that grating monochromators always have a higher luminosity than have prism monochromators at the same bandwidth.

6.29. Double monochromators

A major problem in the design of monochromators is the elimination of scattered light; it is in fact inevitable that some parasitic radiation scattered by the various surfaces in the ray paths will reach the exit slit. This radiation will be of different spectral composition from that of the directly transmitted radiation and is consequently a source of error; it can completely falsify measurements when the desired radiation happens to lie in a spectral region where the source radiance is very low compared with that in other regions or where the detector has a low sensitivity.

The elimination of this source of error calls for a double monochromator—two identical monochromators mounted in series with the exit slit of the first doubling as entrance slit for the second. Two dispositions are possible: in one the separate dispersions are added while in the other they are in opposition.

For any two radiations differing in wavenumber by $\Delta\sigma$ there are two corresponding slit images separated by

$$\Delta x = \frac{dx}{d\sigma}\Delta\sigma$$

$dx/d\sigma = f\, d\theta/d\sigma$ being the linear dispersion. In the first arrangement (Fig. 6.27), the dispersion due to the second monochromator is added to that of the first so that the separation becomes $2\Delta x$ at the exit slit; this arrangement is known as an additive dispersion mounting.

The dispersion due to the second monochromator may, on the other hand, be made to suppress that of the first by adopting

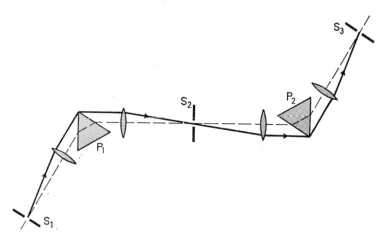

Fig. 6.27. Double monochromator: additive dispersion mounting.

the subtractive dispersion mounting indicated in Fig. 6.28. Here, the monochromatic images separated at the intermediate slit but remaining within its width are recombined by the second monochromator. Both systems are equally effective in eliminating scattered light.

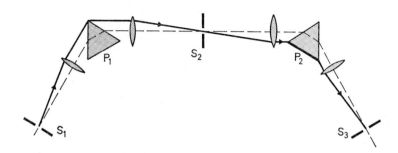

Fig. 6.28. Double monochromator: subtractive dispersion mounting.

SPECTROMETERS AND MONOCHROMATORS

In both cases the best performance is obtained when all three slits have the same angular width. The properties of the two mountings differ only in points of detail.

In the subtractive mounting it is obvious that the only purpose of the second monochromator is to reject the scattered radiation: the bandwidth is not reduced by its presence since all the radiation passing the intermediate slit after dispersion by the first monochromator is focused within the exit slit aperture. The width of the latter is therefore not very critical; flux is lost if it is less than that of the other two while if it is slightly wider the increase of scattered light is negligible. The width of the pass band is therefore defined by the first monochromator and the emergent flux is that passing through the central slit less the transmission loss in the second monochromator.

The foregoing conclusions only apply to the subtractive mounting. When dispersions are additive, the second monochromator plays a part in limiting the bandwidth and, consequently, the transmission of flux. With all three slits of equal width, the bandwidth passed by the second monochromator is half that due to the first: the emergent flux is also reduced to one-half of that reaching the intermediate slit (apart from the transmission losses).† In this case, the adjustment of the exit slit is evidently of greater importance than in the preceding case.

---

† It is a fallacy to suppose that the flux passed by the first monochromator could be wholly transmitted by the second by doubling the width of S_3. If monochromator I transmits a flux Φ with a bandwidth $\Delta\sigma$, the corresponding values for the double monochromator are $\frac{1}{2}\Phi$ and $\frac{1}{2}\Delta\sigma$ when all three slits are of equal width. If the final bandwidth is to remain at $\frac{1}{2}\Delta\sigma$ when S_3 has double the width of S_1 and S_2 it must be reduced to $\frac{1}{4}\Delta\sigma$ at the exit of the first monochromator: the transmitted flux will then be only $\frac{1}{4}\Phi$ since it varies with the square of bandwidth.

CHAPTER SEVEN

INTERFERENCE SPECTROSCOPY

7.1. Introduction

It is not easy to define the limits of interference spectroscopy since, as already pointed out, it can be argued that all dispersing elements, even prisms, depend ultimately on a manifestation of interference. However, some practical basis of classification must be adopted to facilitate an orderly discussion so we conform here to common usage, which excludes prism and grating instruments from the interference class.

In this chapter, then, we shall examine the use in spectroscopy of interferometers such as the Michelson, the Fabry–Perot and the polarization interferometer. A consideration of the principles of these instruments will show that in effect our classification amounts to reserving the name interference spectroscopy for those techniques in which the spectral sorting of the radiations is achieved by the use of interference in the high-order region. Whereas the interference order is generally less than ten for gratings (and zero for prisms), the order in interference spectrometers is rarely less than ten thousand and often exceeds a million.

7.2. Principal methods of interference spectroscopy

Many methods are available for interference spectroscopy and this area of spectroscopy has seen great activity in recent years. In view of the many variations on instrument types, it seems preferable to consider the basic principles employed rather than to attempt to classify each instrument individually.

INTERFERENCE SPECTROSCOPY 145

This leads to a natural division between those instruments that produce a spatial dispersion of the spectrum and those that distinguish radiations of differing frequencies by selectively labelling them by imposed variations in amplitude or modulation frequency, without spatial separation.

The first, spatial dispersion, class is uniquely represented by the *Fabry–Perot étalon spectrometer*; after dealing with this instrument we shall examine the more complex amplitude-modulation and frequency-modulation instruments which serve the technique now commonly known as *Fourier spectroscopy*. This technique, which has mainly been developed in recent years, shows great power and promise.

7.3. Application of an interferometer as a dispersing instrument

We begin by demonstrating that any interferometer can produce a spatial separation of radiations of differing frequencies and is therefore a dispersing device.

Suppose that monochromatic radiation is incident on the interferometer. At any point M in the interference field, the path difference δ between the two interfering beams (in two-beam interferometers) or between two successive interfering beams (in multiple-beam interferometers) is exactly determined; this must be so if a set of fringes exists. At M, then, the order of interference p only varies with the wavenumber σ of the incident radiation, since $p = \sigma \delta$. Any variation $d\sigma$ of σ appears at M as an order variation dp given by $dp/p = d\sigma/\sigma$ and by a displacement dx of the fringes. This displacement is

$$dx = i \, dp = ip \, d\sigma/\sigma$$

since the fringe spacing i corresponds to a unit change in p.

If now the interferometer receives two superimposed radiations of wavenumbers σ and $\sigma + d\sigma$, two fringe systems are formed with a relative displacement dx. The sorting of the two radiations is thus effected by the separation of their respective fringe systems. It is possible to calculate on this basis the theoretical resolving power of an interferometer illuminated by heterochromatic radiation.

7.4. Theoretical resolving power of an interferometer

Two neighbouring interference fringes may be regarded as being visibly separated if their centres are separated by a distance at least equal to their half-width. While this is a somewhat arbitrary criterion, it has the advantage of being simple to apply and it leads to reasonable results, provided that any difference in fringe intensities is not too great.

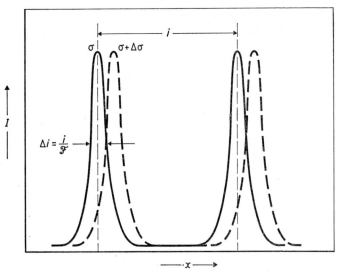

FIG. 7.1. Relative positions of interference fringes due to two radiations differing in wavenumber by $\Delta\sigma = \sigma/p\mathscr{F}$.

The sharpness of a monochromatic set of fringes is characterized by the *coefficient of finesse* $\mathscr{F}$, the ratio of fringe spacing to the half-width of the fringes. Two neighbouring fringes, due to two radiations separated in wavenumber by $\Delta\sigma$, are by definition separable if the distance between their centres is a length Δx at least equal to a fraction $1/\mathscr{F}$ of the fringe spacing in either set (Fig. 7.1). Since $\Delta x/i = p\,\Delta\sigma/\sigma$, the minimum value of $\Delta\sigma$ for which separation is possible is such that $p\,\Delta\sigma/\sigma = 1/\mathscr{F}$ which immediately gives the theoretical resolving power of the instrument

$$\mathscr{R}_0 = \frac{\sigma}{\Delta\sigma} = p\mathscr{F} \qquad (7.1)$$

Since there is no *a priori* reason to limit the order of interference, the theoretical resolving power of an interference spectroscope is, in principle, unlimited: herein lies one of the primary advantages of this type of dispersing instrument. The coefficient of finesse appears at first sight to be only of secondary importance if the order of interference can be made indefinitely high but in practice this is not so, as will be shown when the free spectral range is taken into account.

7.5. Free spectral range: importance of coefficient of finesse

At some point in the interference field where the path difference is δ, bright fringes due to all radiations satisfying the relationship

$$\sigma = k/\delta$$

will be superimposed. k can be any positive integer, so that there will be a series of values of σ forming an arithmetical progression of which the common difference $\Delta\sigma_0 = 1/\delta$ is the free spectral range. As might be expected, this is the same equation for $\Delta\sigma_0$ as that for the case of a grating (see § 4.4). The free spectral range must be as large as possible since it measures the extent of the spectrum that can be accommodated by the instrument without overlapping of spectral orders. In comparing interferometers and gratings, it is useful to compare the values of $\Delta\sigma_0/\Delta\sigma$, the ratio of the free spectral range to the resolved spectral interval. For the interferometer, since $\Delta\sigma_0 = 1/\delta = \sigma/p$, and $\Delta\sigma = \sigma/\mathscr{R}_0$, we have

$$\Delta\sigma_0/\Delta\sigma = \mathscr{R}_0/p = \mathscr{F} \quad (7.2)$$

whereas for the grating $\Delta\sigma_0/\Delta\sigma$ is equal to the total number of rulings. In this respect, the interferometer is inferior to the grating, for while a grating may easily have 100 000 rulings, the finesse of interferometer fringes never exceeds the order of ten; the free spectral range is thus much smaller for the interferometer.

It will later be shown that this disadvantage can be minimized, so making it possible to use interferometers for the analysis of wide-band spectra. This can be achieved without sacrificing the fundamental advantages of the interferometer as a dispersing element, namely, unlimited resolving power and a luminosity much higher than that of grating spectrometers.

A high coefficient of finesse is clearly essential if the interferometer is to be a useful dispersing element. Only a multiple-beam interferometer can, therefore, be considered for this purpose: the Fabry–Perot is the obvious choice, on the grounds of flexibility and ease of construction. This instrument has now completely displaced its forerunners such as the Michelson échelon and the Lummer–Gehrke plate.

THE FABRY–PEROT INTERFEROMETER IN SPECTROSCOPY

7.6. *Description of the Fabry–Perot interferometer*

Construction

The physical basis of the Fabry–Perot interferometer (Fig. 7.2) is a pair of transparent plates P_1 and P_2 (generally of glass or quartz) separated by an air space. The facing surfaces of P_1 and P_2 are ideally perfectly plane and parallel. When a non-collimated beam of radiation passes through the plates the multiple reflections between the two facing surfaces give rise to a set of interference fringes located at infinity; these may be projected onto a screen in the focal plane M of a lens (or viewed in that plane by means of an eyepiece), where they appear as a series of concentric rings centred on a normal to the operative surfaces of the plates. To enhance the sharpness of the fringes, these surfaces are coated with high-reflection films; these may be semi-transparent metal coatings or, better, multi-layer dielectric films. The plates are slightly wedge-shaped so that the fringe system is not disturbed by reflections from the two outer surfaces.

The most common form of this instrument is known as the Fabry–Perot étalon. In this form, the separation of the plates is maintained by three spacers of low-expansion material such as Invar or silica. The spacers are made as nearly equal in thickness as possible and the final adjustment of parallelism is made by applying compressive loads through springs. In this way the two reflecting surfaces can be made as nearly parallel as the defects in their surface form will permit (see § 7.11 below). The setting so obtained is generally very stable.

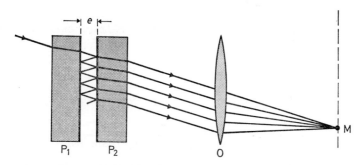

FIG. 7.2. Principle of the Fabry–Perot interferometer.

Optical principles

The surface coatings of the plates will be assumed to be identical. Let their reflection, transmission, and absorption factors be R, T and A respectively. The separation of the surfaces is e and the refractive index of the air space n. We shall now calculate the distribution of intensity in the interference pattern formed in transmission. Successive reflections of an incident ray falling on the interferometer at an angle i give rise to an infinity of transmitted rays all parallel to the directly transmitted ray (Fig. 7.2). The path difference between two successive rays is $\delta = 2ne\cos i$ and hence the phase difference $\phi = 2\pi\sigma\delta = 2\pi p$ where σ is the wavenumber *in vacuo* of the incident radiation and $p = 2ne\sigma\cos i$ is the order of interference. The intensities of the series of transmitted rays are (taking the intensity of the incident ray as unity)

$$T^2, T^2R^2, \ldots, T^2R^{2n}, \ldots$$

The corresponding complex amplitudes are therefore

$$T, TRe^{-j\phi}, \ldots, TR^n e^{-jn\phi}, \ldots$$

These form a geometrical progression with the common ratio $Re^{-j\phi}$.

The resultant complex amplitude of the superposition at M of the infinite series of transmitted components is

$$\mathscr{A} = T(1 + Re^{-j\phi} + \cdots + R^n e^{-jn\phi} + \cdots)$$

$$= \frac{T}{1 - Re^{-j\phi}}$$

Since the intensity of the incident ray is taken as unity, the intensity transmission factor $\mathscr{T}$ of the interferometer is the square of the modulus of $\mathscr{A}$:

$$\mathscr{T} = \mathscr{A}\mathscr{A}^* = \frac{T^2}{|1 - Re^{-j\phi}|^2} = \frac{T^2}{(1 - R\cos\phi)^2 + R^2 \sin^2\phi}$$

$$= \frac{T^2}{1 + R^2 - 2R\cos\phi} = \frac{T^2}{(1-R)^2} \cdot \frac{1}{1 + m \sin^2(\phi/2)}$$

in which $m = 4R/(1-R)^2$.

Putting $\mathscr{T}_M = T^2/(1-R)^2$ for the maximum value of $\mathscr{T}$, the expression becomes

$$\mathscr{T} = \frac{\mathscr{T}_M}{1 + m \sin^2(\pi p)} \qquad (7.3)$$

If monochromatic radiation is being transmitted, the order of interference and, in consequence, the transmission factor $\mathscr{T}$, depend only on the angle of incidence i; the result is the familiar system of ring fringes. The characteristics of the transmission function $\mathscr{T}$ will be investigated later.

7.7. Various methods of application of the Fabry–Perot étalon as a dispersing instrument

When the spectral content of the incident radiation is no longer monochromatic, dispersion is apparent as a spatial separation of the fringe systems due to the radiations of differing wavelengths. In other words the ring diameter depends on wavelength. The spectrum thus obtained may be recorded either as a photographic image or as the output of a flux detector.

The photographic method is the older and has been widely used since the invention of the Fabry–Perot interferometer itself in 1897, mainly in the study of the hyperfine structure of spectral lines. Not only is this method still in use but, as we shall see, a recent refinement has given it a new lease of life.

The use of a flux detector (usually a photomultiplier) is much more recent, having been introduced by P. Jacquinot and C. Dufour in 1948.[51] The photoelectric method has, however, made great progress and is now widely used; we shall therefore give first attention to this method.

FABRY–PEROT ÉTALON SPECTROMETER

7.8. Basic construction of a Fabry–Perot étalon spectrometer

As the output is to be analysed by a flux detector, a slit must be provided to scan the spectrum, as in the case of prism and grating spectrometers, but here the slit has to be in the form of a circular annulus. In practice it is more convenient simply to use a diaphragm with a circular hole centred on the ring fringe system. Scanning of the spectrum is in this case effected by varying the optical thickness ne of the air space between the interferometer plates.

The arrangement of the spectrometer components is shown in Fig. 7.3.

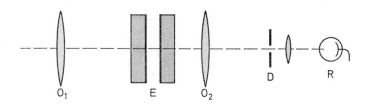

Fig. 7.3. Diagram of Fabry–Perot étalon spectrometer.

The radiation to be analysed is collimated by lens O_1. The interference rings are formed in the focal plane of lens O_2, where the diaphragm D is located. Immediately behind D is the photomultiplier or other detector R. Provision is made for continuous variation of the separation of the étalon plates, thus generating a continuous scan of the spectral distribution of fringes as seen by the detector through the diaphragm.

7.9. Factors contributing to the instrumental profile of the spectrometer

In § 6.2 the instrumental profile was defined as that represented by the curve recorded when the instrument is scanned through a spectral region containing only a single strictly monochromatic radiation; this is equivalent to recording the output while the instrument setting remains fixed and the wavenumber of the incident radiation is continuously increased. The instrumental profile describes the total effect of all the factors that cause a broadening of an infinitely narrow spectral line and, as we have already seen, it can be represented by the convolution product of several partial profiles, each being the result of a particular cause of line broadening.

For the Fabry–Perot spectrometer there are three different causes of broadening to be taken into account.[6] First of all, the fringes themselves have a limited finesse, that is, their width can never be zero. This theoretical finesse depends only on the reflectivity of the plates and is consequently termed the reflective finesse $\mathscr{F}_R$. In practice, the effective finesse is also limited by the inevitable imperfections of the reflecting surfaces of the plates which give rise to local variations of the étalon spacing. Finally, allowance must be made for the finite diameter of the diaphragm aperture.

We now have to derive three partial instrumental profiles $A_1(\sigma)$, $A_2(\sigma)$ and $A_3(\sigma)$, corresponding to these three effects, in order that the total instrumental profile may be calculated.

7.10. Influence of the reflectivity of the plates: profile $A_1(\sigma)$

A Fabry–Perot étalon of strictly constant spacing has a transmission factor $\mathscr{T}$ for monochromatic radiation of wavenumber δ given by the equation

$$\mathscr{T} = \frac{\mathscr{T}_M}{1 + m \sin^2 \pi \sigma \delta} \qquad (7.4)$$

which defines the Airy function. Different values of this function would be recorded as the étalon spacing changes if the focal plane aperture were infinitely small. The instrumental profile of an ideal Fabry–Perot spectrometer is therefore expressed by the Airy function (Fig. 7.4); it consists of a series of sharply defined maxima separated by the free spectral range which, as was shown in § 7.5, has the value $1/\delta$.

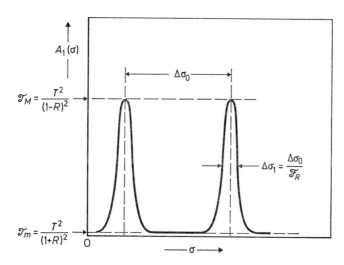

FIG. 7.4. Airy function $A_1(\sigma)$.

The half-width $\Delta\sigma_1$ of the maxima of the Airy curve is calculated as follows. If σ_k is the abscissa of a maximum, we have

$$\mathscr{T} = \tfrac{1}{2}\mathscr{T}_M \quad \text{for} \quad \sigma = \sigma_k \pm \tfrac{1}{2}\Delta\sigma_1$$

hence

$$\sin^2\{\pi(\sigma_k \pm \tfrac{1}{2}\Delta\sigma_1)\delta\} = 1/m$$

But, since σ_k refers to a maximum, the corresponding order of interference $\sigma_k\delta$ is an integer and the preceding equation becomes:

$$\sin^2(\tfrac{1}{2}\pi\delta\,\Delta\sigma_1) = 1/m = \sin^2(\tfrac{1}{2}\pi\,\Delta\sigma_1/\Delta\sigma_0) \qquad (7.5)$$

It can be assumed that the fringe finesse is sufficiently high, and hence the ratio $\Delta\sigma_1/\Delta\sigma_0$ small enough, for equation (7.5) to be written

$$\tfrac{1}{2}\pi\, \Delta\sigma_1/\Delta\sigma_0 = m^{-1/2}$$

Hence the half-width is

$$\Delta\sigma_1 = 2\,\Delta\sigma_0/\pi\sqrt{m} = (1-R)\,\Delta\sigma_0/\pi\sqrt{R}$$

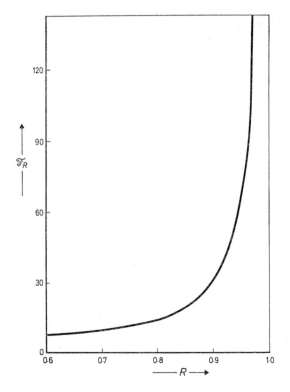

FIG. 7.5. Variation of coefficient of reflective finesse $\mathscr{F}_R = \pi\sqrt{R}/(1-R)$ of fringes formed by an ideal Fabry–Perot étalon as a function of the reflectivity R of the plates.

The ratio $\Delta\sigma_0/\Delta\sigma_1 = \pi\sqrt{R}/(1-R)$ is simply the theoretical finesse of the fringes, or reflective finesse $\mathscr{F}_R$. Fig. 7.5 shows the variation of this quantity with the reflectivity R of the plates.

INTERFERENCE SPECTROSCOPY

Conversely, the maximum transmission factor $\mathcal{T}_M$ is theoretically independent of R if the absorption factor A for the reflecting coatings is negligible; in fact

$$\mathcal{T}_M = T^2/(1-R)^2 = 1 \quad \text{if } A = 1-T-R = 0$$

It would appear that under these conditions the highest possible value of R should be applied in order to maximize the reflective finesse. We shall see later, however, that finesse is eventually limited by the surface form of the plates and that it is useless to raise $\mathcal{F}_R$ above a certain limit.

7.11. *Influence of surface form of the plates: function $A_2(\sigma)$*

In deriving the instrumental profile $A_1(\sigma)$, the spacing e of the interferometer was assumed to be uniform. In fact, it is not yet possible to produce surfaces of which the effect on e of departures from flatness can be neglected. This is a fact that can easily be demonstrated by experiment: it is only necessary to put a Fabry–Perot étalon in a collimated monochromatic beam of wavelength such that the nominal plate separation e_0 gives maximum transmission. If the separation were strictly constant over the whole area of the plates the illuminance on a screen placed immediately behind the étalon would be uniform. What is in practice observed is an irregular distribution of bright patches, showing that the correct thickness only occurs over certain zones. If the wavelength of the radiation or the separation of the plates is varied, the bright zones shift and change their shape. These variations obviously indicate the presence of defects of flatness on the reflecting surfaces.

R. Chabbal, following the work of Dufour and Picca,[35] undertook a detailed examination of the influence of these defects on the instrumental profile of the spectrometer.[6,10] He has shown that an imperfect étalon can be regarded as being composed of a mosaic of small perfect étalons of differing separations, the total transmitted flux being the sum of the transmissions of all the elementary étalons. If e_0 is the mean separation of the real étalon, the separation at some particular point is $e = e_0 + \varepsilon$ (Fig. 7.6); let dS be the total area of the zones for which e lies between $e_0+\varepsilon$ and $e_0+\varepsilon+d\varepsilon$. The elementary étalon of area dS transmits a flux

$$d\Phi = \frac{\mathcal{T}_M}{1+m\sin^2\{2\pi n(e_0+\varepsilon)\sigma\cos i\}}\,dS$$

The surface deformation of the plates can be characterized[10] by the function

$$\mathscr{D}(\varepsilon) = dS/d\varepsilon$$

which can be evaluated for some typical surfaces. By putting

$$\sigma' = -\varepsilon\sigma/e_0 \tag{7.6}$$

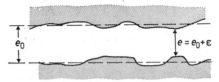

FIG. 7.6. Imperfect Fabry–Perot étalon.

and substituting the new expression for ε given by equation (7.6) in $\mathscr{D}(\varepsilon)$, we have $dS = \mathscr{D}(\varepsilon)\,d\varepsilon = D(\sigma')\,d\sigma'$ and hence

$$d\Phi = \frac{\mathcal{T}_M}{1+m\sin^2\{2\pi n e_0(\sigma-\sigma')\cos i\}}\,D(\sigma')\,d\sigma'$$

The total flux Φ transmitted by the real étalon is the sum of the fluxes transmitted by the elementary components, so

$$\Phi(\sigma) = \int \frac{\mathcal{T}_M}{1+m\sin^2\{2\pi n e_0(\sigma-\sigma')\cos i\}}\,D(\sigma')\,d\sigma' \tag{7.7}$$

or

$$\Phi(\sigma) = A_1(\sigma) * D(\sigma)$$

This expression for the flux Φ as a function of wavenumber represents the instrumental profile of an imperfect Fabry–Perot étalon. Equation (7.7) shows Φ as the convolution of the instrumental profile $A_1(\sigma)$ of an ideal instrument and the function $D(\sigma')$. This latter function must therefore be the instrumental profile of the elementary étalons, determined by the characteristics of the surfaces. From now on we can put $D(\sigma') = A_2(\sigma')$ and hence for the real étalon

$$\Phi(\sigma) = A_1(\sigma) * A_2(\sigma') = A_{12}(\sigma)$$

Since σ' is by definition proportional to ε, the functions $D(\sigma')$ and $A_2(\sigma')$ have the same form; they differ only in scale.

It is particularly important to know the width $\Delta\sigma_2$ of the function $A_2(\sigma')$. This is, of course, proportional to the width $\Delta\varepsilon$ of $\mathscr{D}(\varepsilon)$, the relationship being

$$\Delta\sigma_2 = \frac{\sigma}{e_0}\Delta\varepsilon$$

Introducing the free spectral range $\Delta\sigma_0$ enables us to write

$$\frac{\Delta\sigma_0}{\Delta\sigma_2} = \frac{e_0 \Delta\sigma_0}{\sigma \Delta\varepsilon}$$

but

$$\Delta\sigma_0 = \frac{1}{2ne_0 \cos i} \approx \frac{1}{2e_0}$$

giving

$$\frac{\Delta\sigma_0}{\Delta\sigma_2} = \frac{\lambda}{2\Delta\varepsilon} = \mathscr{F}_D \qquad (7.8)$$

The ratio $\Delta\sigma_0/\Delta\sigma_2$ has the form of a coefficient of finesse; it can be termed the defect finesse since if the width $\Delta\sigma_1$ of the Airy function maxima is small compared with $\Delta\sigma_2$, that is, if $\mathscr{F}_R \gg \mathscr{F}_D$, this second coefficient is operative in determining the finesse of the interference fringes or the maxima of the instrumental profile of the interferometer.

It should be noted that $\mathscr{F}_D$ is a very simple function of $\Delta\varepsilon$ and depends, not on the étalon spacing, but only on the variations in the spacing or, in other words, on surface imperfections and incidentally on errors in adjustment of the plates.

The function $\mathscr{D}(\varepsilon)$ can be identified for certain specifiable surface deformations. It can be shown that, for example, it is a rectangular function for a spherical surface and a Gaussian function for a random microstructure. In practice, surfaces seldom have defects of such simple forms alone but the shape of the function $\mathscr{D}(\varepsilon)$ is of little importance: what matters is its width or the associated finesse $\mathscr{F}_D$; we shall show that it is possible to measure these quantities directly.

7.12. Influence of the dimensions of the diaphragm aperture: function $A_3(\sigma)$

The instrumental profile $A_3(\sigma)$ representing the influence of the aperture will clearly have a rectangular form as in the case of prism or grating spectrometers. This is easily seen by considering the change of transmitted flux as the spectrum is scanned, supposing the fringes to be very much narrower than the diameter of the aperture: while a fringe is expanding within the area of the aperture the flux remains constant but it falls suddenly to zero as the ring expands beyond the boundary of the aperture. The width $\Delta\sigma_3$ of $A_3(\sigma)$ may be related to the geometrical width of the aperture by considering the scanning of the aperture by a fringe. Let $\Delta(\cos i)$ be the variation of $\cos i$ required to scan the aperture (bearing in mind that the aperture is to be regarded as an annular slit of zero inner radius).

Since the order of interference $p = 2ne\sigma \cos i$ is constant for any one fringe the variation in wavenumber $\Delta\sigma$ required to scan the aperture is such that $\Delta\sigma/\sigma = \Delta(\cos i)/\cos i$. As $\cos i$ is always nearly unity we can write

$$\Delta\sigma_3 = \sigma \, \Delta(\cos i)$$

This equation indicates the diameter of an aperture corresponding to any given width $\Delta\sigma_3$ of the instrumental profile A_3. The angular radius α of the aperture is obtained from the equation

$$\Delta(\cos i) = 1 - \cos\alpha \approx \alpha^2/2 = \Delta\sigma_3/\sigma$$

or

$$\alpha = \sqrt{(2 \, \Delta\sigma_3/\sigma)} \tag{7.9}$$

7.13. Inclusive instrumental profile

This is the convolution product $A = A_1 * A_2 * A_3$ of the three component profiles. This function $A(\sigma)$ naturally follows

$A_1(\sigma)$ in having a series of equidistant maxima which are separated by the free spectral range $\Delta\sigma_0$ (Fig. 7.7).

The two important characteristics of $A(\sigma)$ are, firstly, the ordinate τ of the maxima and, secondly, the value of $\Delta\sigma$, the half-width. The luminosity of the spectrometer is in fact proportional to τ while its effective resolving power $\mathscr{R}$ varies in inverse ratio to $\Delta\sigma$, since $\mathscr{R} = \sigma/\Delta\sigma$.

Note that the resolving power can theoretically be increased without limit because the width of each of the component profiles can be made as small as desired. This is obviously true for

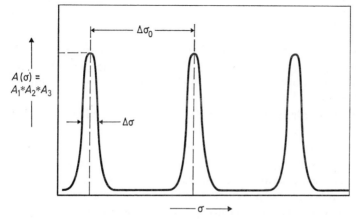

FIG. 7.7. Instrumental profile of Fabry–Perot étalon spectrometer.

A_3, which diminishes with the aperture diameter, but is also true for A_1 and A_2, since $\Delta\sigma_1 = \Delta\sigma_0/\mathscr{F}_R$ and $\Delta\sigma_2 = \Delta\sigma_0/\mathscr{F}_D$.

Now, the free spectral range $\Delta\sigma_0$ is reduced simply by increasing the separation of the plates; it follows that by this means $\Delta\sigma_1$ and $\Delta\sigma_2$, and hence the resultant width $\Delta\sigma$, are reducible at will. Even the *effective* resolving power $\mathscr{R}$ has no fundamental limit. This represents one of the primary advantages of interference spectroscopy.

An exact knowledge of the instrumental profile $A(\sigma)$ would require a knowledge of each of the three constituents. As already indicated, it is impossible to state the exact form of the profile A_2 for the general case of surface deformation; only its width $\Delta\sigma_2$ can be known precisely. Fortunately, the significant parameters—τ the height and $\Delta\sigma$ the width of the maxima of the inclusive profile A—depend much less on the exact shape of $A_2(\sigma)$ than on its width; this enables valid magnitudes of those parameters to be calculated for all cases. A useful indication is that $\Delta\sigma$ is always larger than the greatest of the widths $\Delta\sigma_1$, $\Delta\sigma_2$ and $\Delta\sigma_3$ but less than their sum.

The remaining problem is to choose the instrument parameters in such a way as to yield the most favourable instrumental profile, that is, one that gives the best combination of resolving power and luminosity.

7.14. *Importance of surface defects of the plates*

The least possible width of the inclusive instrumental profile is necessarily set by the largest of the three values $\Delta\sigma_1$, $\Delta\sigma_2$, $\Delta\sigma_3$, so that we must first know which is liable to have the greater influence.

We have already seen that $\Delta\sigma_3$ can be chosen to be as small as is desired by reducing the scanning aperture. Equally the reflective finesse can be made almost indefinitely high, at least in the visible region, by virtue of the very high reflectivities obtainable with surface coatings (values in the region of 0·998 in the red have been reported for surface-coated laser mirrors). On the other hand, it is not possible to work étalon plates to better than $\lambda/100$ over a sufficiently large area; this, having regard to the combined effect of the two plates, means that according to equation (7.8) the defect finesse $\mathscr{F}_D$ is limited to a value in the region of, say, 40–50. It is therefore the profile A_2 due to surface defects that is the widest of the three. Thus an inclusive (or effective) finesse $\mathscr{F}$ can be ascribed to the spectrometer, defined as $\mathscr{F} = \Delta\sigma_0/\Delta\sigma$, the ratio of the free spectral range to the width of the maxima of the instrumental profile $A(\sigma)$. From the foregoing discussion it will be clear that $\mathscr{F}$ is always smaller than the $\mathscr{F}_D$ ascribed to defects. The latter acts as the limiting finesse.

The problem before us is, therefore, the choice of values for $\Delta\sigma_1$ and $\Delta\sigma_3$ (that is, for reflective finesse $\mathscr{F}_R$ and angular radius α of the scanning aperture) which, when taken in conjunction with the limiting defect finesse $\mathscr{F}_D$, result in the optimum value for the instrumental profile $A(\sigma)$.

7.15. Choice of reflective finesse and of radius of scanning aperture[6,7,10]

The choice of reflectivity of the plates is approached by first considering the partial instrumental profile

$$A_{12}(\sigma) = A_1 * A_2$$

This may be regarded as the instrumental function of the real étalon alone, without the scanning aperture. Its shape is similar to that of the profiles A_1 and A (Figs. 7.4 and 7.7). The width $\Delta\sigma_{12}$ of its maxima can be represented by a new coefficient of

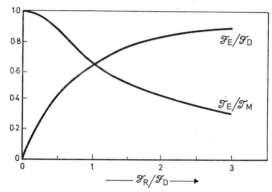

FIG. 7.8. Variation of the maximum transmission $\mathscr{T}_E$ of a real étalon and of the finesse $\mathscr{F}_E$ of the corresponding fringes as a function of the ratio $\mathscr{F}_R/\mathscr{F}_D$ of the reflective finesse to the limiting defect finesse.

finesse $\mathscr{F}_E = \Delta\sigma_0/\Delta\sigma_{12}$, the étalon finesse, which is simply the effective finesse of the interference fringes. The ordinate $\mathscr{T}_E$ of the maxima of A_{12} gives the maximum transmission factor of the practical étalon.

The variation of $\Delta\sigma_{12}$ (or $\mathscr{F}_E$) and $\mathscr{T}_E$ as a function of the reflective finesse must now be found.

The curves giving these variations are shown in Fig. 7.8.†

† The exact shape of these curves depends on the nature of the function $A_2(\sigma)$. The curve shown in Fig. 7.8 is a curve typical of those resulting from the most commonly occurring values of $A_2(\sigma)$. For a more detailed study, see references 6 and 10.

Here the abscissae are in terms of the ratio $\mathscr{F}_R/\mathscr{F}_D$, in order to take account of the fact that $\mathscr{F}_D$ has a predetermined value. The ordinates for the reflectivity curve are normalized in the same way and for the transmissivity curve the normalizing factor is $\mathscr{T}_M = T^2(1-R)^2$, the value for a perfect étalon.

Fig. 7.8 shows very clearly that there is no point in raising the reflective finesse much above $\mathscr{F}_D$: as $\mathscr{F}_R/\mathscr{F}_D$ rises above 1, the étalon finesse $\mathscr{F}_E$ does not increase very much but the transmission factor $\mathscr{T}_E$ falls off rapidly.

A value for the ratio $\mathscr{F}_R/\mathscr{F}_D$ significantly higher than unity thus brings a serious loss of luminosity without a useful gain of resolving power.

The optimum value of $\mathscr{F}_R$ therefore lies near to $\mathscr{F}_D$. An examination of the optimum value of $\Delta\sigma_3$ for the scanning aperture would lead to a similar conclusion.

The conclusion is that to maintain the highest possible luminosity compatible with a given resolving power the partial profile widths $\Delta\sigma_1$, $\Delta\sigma_2$ and $\Delta\sigma_3$ must have approximately equal values. The surface quality thus determines the optimum width of the instrumental profile or free spectral range for the étalon, the relationship being $\Delta\sigma \approx \Delta\sigma_2/0{\cdot}6$. The effective resolving power is $\mathscr{R} = 0{\cdot}6\mathscr{R}_0$, $\mathscr{R}_0$ being an idealized resolving power corresponding to zero widths for the profiles A_1 and A_3.

The angular radius α (see § 7.12) of the scanning aperture should therefore be given the value

$$\alpha = \sqrt{(2\,\Delta\sigma_3/\sigma)} = \sqrt{(2/\mathscr{R}_0)}$$

since $\Delta\sigma_3 = \Delta\sigma_2$.

Under these conditions, the maximum value τ of the instrumental profile $A(\sigma)$ which controls the luminosity of the spectrometer is in the region $\tau = 0{\cdot}6\mathscr{T}_M$, $\mathscr{T}_M$ being the maximum value of the Airy function or, in other words, the maximum transmission of the equivalent perfect étalon.

7.16. *Evaluation of the luminosity*

The definition of étalon luminosity is, as for the slit spectrometer, the ratio $\mathscr{L} = \Phi/L$ (Φ, flux transmitted by the scanning aperture; L, radiance of source). The flux is in the present case given by the product

$$\Phi = L\tau S\Omega$$

in which τ is the transmission factor as defined above, S the effective area of the objective lens (in practice that of the étalon) and Ω the solid angle subtended by the scanning aperture A at the focusing lens O_2 (Fig. 7.9). We have

$$\tau = 0.6 \frac{T^2}{(1-R)^2}, \quad \Omega \approx \pi\alpha^2 = \frac{2\pi}{\mathscr{R}_0}$$

so that

$$\mathscr{L} = \tau SQ \approx 3.6 \frac{T^2}{(1-R)^2} \frac{S}{\mathscr{R}_0} \quad (7.11)$$

or, in terms of the effective resolving power $\mathscr{R} \approx 0.6\mathscr{R}_0$,

$$\mathscr{L} = 2.2 \frac{T^2}{(1-R)^2} \frac{S}{\mathscr{R}}$$

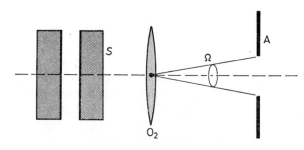

FIG. 7.9. Evaluation of étendue of beam transmitted by Fabry–Perot spectrometer.

With multilayer dielectric coatings (to be described later) values for $T^2/(1-R)^2$ range between 0.5 and 0.9. Taking a mean value of 0.7:

$$\mathscr{L} = 1.5 S/\mathscr{R} \quad (7.12)$$

As might be expected, we again find a reciprocity between luminosity and resolving power. The invariant $\mathscr{LR}$ depends only on the usable area of the étalon. Although in theory there

164   SPECTROSCOPY AND ITS INSTRUMENTATION

is no limit to the resolving power, a practical limit is imposed by the loss of luminosity accompanying an increase in resolving power.

7.17. *Comparison with the grating spectrometer*[53]

In § 6.10, the product $\mathscr{LR}$ for a slit spectrometer was expressed in terms of the transmission factor τ, S the cross-sectional area of the beam leaving the dispersing element, β the angular height of the slits and $d\theta/d\sigma$ the angular dispersion:

$$\mathscr{LR} = \tau S \beta \sigma \, d\theta/d\sigma$$

For grating spectrometers† the angular dispersion may be calculated from the basic equation (see § 4.2)

$$\sin \theta_2 - \sin \theta_1 = k\lambda/c$$

We have

$$\sigma \left|\frac{d\theta_2}{d\sigma}\right| = \lambda \left|\frac{d\theta_2}{d\lambda}\right| = \frac{|k|\lambda}{c \cos \theta_2} = \frac{|\sin \theta_2 - \sin \theta_1|}{\cos \theta_2}$$

Also, the area A of the grating is equal to $S/\cos \theta_2$ so that

$$\mathscr{LR} = \tau A \beta |\sin \theta_2 - \sin \theta_1|$$

It must be remembered that the expression for $\mathscr{LR}$ used here is only applicable in the case where the effective resolving power is considerably smaller than the intrinsic resolving power of the grating: for the intrinsic case $\mathscr{LR}$ must be multiplied by a factor K, lying between 0 and 1, of which the appropriate values are indicated in Fig. 6.5, p. 106.

For simplicity, K will be taken as unity. Also, for the Littrow class of mounting, $\sin \theta_2 - \sin \theta_1$ can be simplified to $2 \sin \phi$, ϕ being the blaze angle and typically having a value of 30° so that $\sin \theta_2 - \sin \theta_1$ will equal 1. For these conditions, then,

$$\mathscr{LR} = \tau A \beta$$

The transmission factor τ is very variable but an average

---

† Prism instruments need not be considered here, their performance being always inferior to that of gratings.

value would be 0·4. We thus arrive at a ratio of the values $\mathscr{L}\mathscr{R}$ for Fabry–Perot and grating spectrometers:

$$\frac{\mathscr{L}\mathscr{R}_{\mathrm{FP}}}{\mathscr{L}\mathscr{R}_{\mathrm{G}}} = \frac{4S}{\beta A}$$

The angular height β of the slits of a grating spectrometer will generally lie within the range $1/100$–$1/10$, giving the above ratio a range of values between $40S/A$ and $400S/A$.

It is unfortunately not possible to make good étalon plates of an area matching that of the largest diffraction gratings but the advantage in luminosity (at a given resolving power) held by the Fabry–Perot over the grating spectrometer is nevertheless considerable.[53]

Other spectrometers, such as the SISAM and the Fourier transform instrument to be described later, share with the Fabry–Perot the advantage of an angular aperture $\Omega = 2\pi/\mathscr{R}_0$ which is much greater than that of the slit spectrometers. P. Jacquinot has shown that basically this superiority resides in the facts that interference spectrometers are symmetrical about the optical axis and that the interfering beams are separated by division of intensity, rather than by spatial division—attributes that have been collectively christened the Jacquinot advantage.

The two fundamental advantages possessed by Fabry–Perot spectrometers are, therefore, a theoretically unlimited resolving power and, for equivalent resolving power, a higher luminosity than that of grating spectrometers. On the other hand, their free spectral range is very limited: it cannot exceed the resolved spectral interval $\Delta\sigma$ by much more than one order. In other words the spectral content of the incident radiation must be restricted to a range less than two orders greater than the resolved interval. This almost always calls for the intervention between source and étalon of a monochromator having a narrow pass band but not restricting the angular aperture of the beam entering the étalon. These conditions are not always easy to satisfy.

7.18. *Elimination of parasitic transmission bands*

If an interference spectrometer could be designed to have a single pass band, then the range and complexity of the spectra that could be examined would impose no restrictions arising from the overlapping of orders; such an instrument could be classified as universal since it would be capable of analysing all types of emission and absorption spectra.

By mounting a monochromator in front of the étalon, it would of course be possible to eliminate all but one of the transmission maxima of a Fabry–Perot spectrometer. The instrumental profile of the monochromator being triangular, it is only necessary to make its half-width not greater than the spectral interval $\Delta\sigma_0$ between successive étalon transmission peaks, to achieve the desired isolation of one pass band (Fig. 7.10).

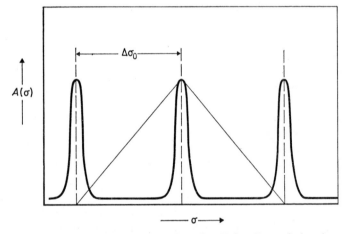

FIG. 7.10. Isolation of one order of a Fabry–Perot étalon by a grating monochromator.

The disadvantage of this method is that $\Delta\sigma_0$ is generally quite small, so that the monochromator must have a high effective resolving power; this demands small slit widths with the consequence of low luminosity and an étendue much smaller than that which the étalon is capable of accepting. This loss of luminosity is made worse by the fact that the pupils of the two instruments are of different shapes. For example, in one typical arrangement, the exit slit of the monochromator would be imaged on the étalon plates; the elongated rectangle of the slit is a poor match to the circular plates and represents a very inefficient coupling with consequent wastage of flux. A grating monochromator is thus not generally useful for this purpose.

There is, however, one very important special case for which the grating method of filtering is often used. In the study of the hyperfine structure of emission lines, the unlimited resolving power and very high luminosity of the Fabry–Perot spectrometer make it the ideal instrument for the purpose. It so happens that the spectral line to be examined is often separated from its neighbours by a comparatively wide gap, so that the bandwidth of the grating monochromator can be allowed a much greater value than $\Delta\sigma_0$ and the luminosity thus raised to an adequately high level.

The more general case of a universal spectrometer calls for a monochromator that has a higher luminosity than a grating instrument and of which the aperture shape matches that of the Fabry–Perot étalon. The natural solution is to use auxiliary étalons of lower spacing than that of the principal étalon. Suppose two étalons of spacings e/k and e (k being an integer) to be mounted in series. The transmission factor of the assembly is the product of the transmission factors of the separate étalons, the resultant instrumental profile having the shape shown in Fig. 7.11.

The half-width of the peaks of the instrumental profile is seen to be slightly less than that of the principal étalon, so that the resolving power is determined by the latter but the introduction of the auxiliary étalon, with its more widely spaced peaks, has multiplied the free spectral range and the finesse by the factor k. Furthermore this type of monochromator imposes less constraint on the beam width than does the principal étalon since its spacing is smaller. The grating monochromator has only one advantage over the auxiliary étalon: it completely suppresses the unwanted Fabry–Perot peaks whereas the transmission factor of the auxiliary étalon is not zero between peaks. The residual transmission, especially that at the wings of the auxiliary étalon peaks, forbids the use of large values of k. For extended free spectral ranges it is therefore necessary to set up a series of étalons of differing spacings, chosen so that the peaks of each étalon coincide at the desired wavelength only (and for this condition k is not necessarily an integer for any pair). An example of a 'universal' spectrometer is given later (see § 7.21) in which the single pass band is achieved with a combination of an interference filter and three Fabry–Perot étalons.

Some studies of hyperfine line structure may call for such high resolving powers that the free spectral range does not even cover the line structure—here again two étalons of different spacings have been used.[14, 15, 16, 59]

The SISAM interferometer due to P. Connes, to be described later (see § 7.27), is another type of interference spectrometer ideally suited to act as a monochromator for an infra-red Fabry–Perot spectrometer.

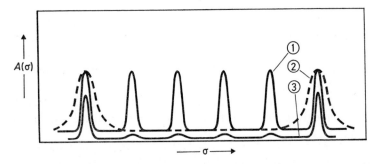

FIG. 7.11. Instrumental profile of an assembly of two Fabry–Perot étalons in series. Ratio of separations $k = 5$.

7.19. *Spectrum scanning*

It is fairly obvious that the spectrum can only be conveniently scanned by causing the interference fringes to be developed in succession in the aperture of a fixed diaphragm. (It must be borne in mind that this scanning aperture is at the centre of the system of ring fringes so that each ring is scanned in its initial phase of formation or final disappearance as a small bright disk.) Scanning is induced by a continuous variation of the optical thickness ne of the interferometer. Variation of either the refractive index n of the gas between the plates or the separation e of the plates is possible; both methods are used in practice.

Variation of plate separation e

Various systems are available for displacing one of the plates with respect to the other but to be acceptable the chosen method must be capable of keeping the plates parallel to a high degree of accuracy. In order that the defect finesse $\mathscr{F}_D$ may be maintained without loss, the parallelism error must remain small compared with departures from surface flatness; this imposes a limit of, say, $\lambda/100$ or 50 Å in the visible region. Fortunately, it is only necessary to maintain this accuracy for a displacement of $\lambda/2$ in scanning one interference order. Even so, automatic correction of errors of parallelism during a scan is sometimes found necessary.

A purely mechanical displacement system is feasible and has been shown to be capable of excellent performance.[3, 9, 60] One of the plates is mounted on elastic hinges carefully designed to impose precisely controlled restraints on the movement.

The use of supports fabricated in a piezo-electric material such as barium titanate ceramic has also been fairly widely adopted. The electrical control of spacing which this system affords offers certain advantages such as the possibility of rapid scanning and of synchronization with some other series of events.[37,67] The magnetostrictive effect can be used in the same way. These two systems have been adopted in particular for Fabry–Perot étalons carried in spacecraft, where their inherent rigidity and compactness are especially appropriate. Plate 7 illustrates an instrument of the magnetostrictive type.[47, 70]

Automatic monitor for parallelism

Whatever the displacement system employed, it is desirable to be able to check the plates for parallelism during the sweep and, if possible, to correct any error automatically. This requirement can be fulfilled by a comparatively simple device (Fig. 7.12).[60, 67, 68]

A beam of white light passes through the étalon near its edge at A, is deflected by prisms B and C and returned through the étalon at a diametrically opposite point D. As far as this beam is concerned the effect is of two étalons in series and a so-called superposition fringe system is set up. The flux transmitted by this system is at a maximum when these two étalons are of exactly equal spacing at A and D but falls sharply as soon as the equality is lost. The slightest tilt of the plates during a scan is thus detectable. To apply automatic correction the intensity of the monitoring beam is measured by a photomultiplier and servo-system operating on the appropriate plate separators. An identical system operates through points on a diameter perpendicular to AD and so ensures parallelism of the plates.

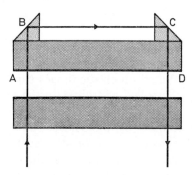

FIG. 7.12. Optical monitor for parallelism of Fabry–Perot étalon plates, using superimposed fringes.

Scanning by change of refractive index

Scanning can be produced quite simply by mounting the étalon inside a pressure chamber and varying the pressure, and hence the refractive index, of the gas. For a scan covering one complete order the path difference $2(n_2 - n_1)e$ must be at least equal to λ; note that the pressure variation required rises as e falls. The relationship between refractive index and pressure is:

$$n_2 - n_1 = C(P_2 - P_1)$$

C is a constant depending on the gas; its value for air is 3×10^{-4} per atmosphere. A pressure change of 1 atmosphere of air therefore induces a change in path difference of $6 \times 10^{-4}e$, which amounts to 6000 Å for $e = 1$ mm. Thus when the spacing e is high enough, as it is in étalons designed for high resolution, the spectrum can be scanned very simply by evacuating the chamber containing the étalon and then admitting air at a slow rate. Various methods have been devised to produce a variation of pressure that is linear with time.[1,16,59] When the étalon spacing is too small to allow for a sufficiently wide variation of refractive index by this simple procedure, air may be replaced by a more highly refractive gas and pressures above one atmosphere employed.

7.20. Practical features of the Fabry–Perot étalon

Materials for the plates

The plate material must of course be transparent in the spectral region to be examined but it is equally important that it be suitable for working to a high degree of flatness and polish. In the ultra-violet, visible and near infra-red regions, fused silica is ideal and is the only material used. Silica becomes absorbent above 2 to 3 μm, in which case the crystalline substances used for infra-red prisms, such as fluorite (CaF_2) and sodium chloride, are available.

High-reflection surface coatings

The working faces of the plates have to be coated with a layer having a reflectivity high enough to give a reflective finesse of the desired value (see Fig. 7.5) but also having as low an absorption factor as possible.

The maximum value of the Airy function (equation 7.3) is

$$\mathscr{T}_M = \frac{T^2}{(1-R)^2} = \frac{T^2}{(T+A)^2} = \frac{1}{(1+A/T)^2}$$

T and A being the transmission factor and absorption factor, respectively, of the coatings.

Since the reflective finesse $\mathscr{F}_R$ is a function of R only ($\mathscr{F}_R = \pi\sqrt{R}/(1-R)$) there is a relationship between $\mathscr{T}_M$ and $\mathscr{F}_R$ for each value of A, as shown in Fig. 7.13.[36,41]

The maximum value of $\mathscr{T}_M$ only approaches unity so long as A is small compared with T or $1-R$. In the past, this condition could not be met because the only known high-reflection coatings were metallic (silver or sometimes, for the ultra-violet, aluminium): all metallic films have a comparatively high absorption factor and the ratio A/T increases with film thickness. This makes it impossible to satisfy the condition of a high étalon transmission with a sufficiently high finesse.

The recent development of high-reflection dielectric multilayer coatings has now made it possible to produce reflecting surfaces having a reflection factor as high as may be desired

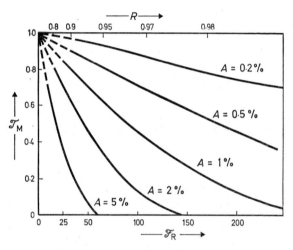

Fig. 7.13. Influence of absorption factor A of reflecting coatings on the characteristics of a Fabry–Perot étalon.

without introducing significant absorption, at least in the visible spectrum. These coatings are of alternate layers of high and low refractive index, the difference between the two indices being as great as possible; there is always an odd number of layers, each of optical thickness $\lambda/4$ so that the reflected components from each interface are all in phase and the resultant reflectivity maximal.

Practical multilayer coatings are usually made with zinc sulphide ($n \approx 2\cdot 3$) and cryolite ($n \approx 1\cdot 3$) (Fig. 7.14). A sufficiently high reflective finesse can be obtained with 7, 5 or even 3 layers. It is quite easy to achieve much higher reflective finesse values than those normally used without inducing significant absorption but a practical limit is set by the surface defects of the étalon plates.

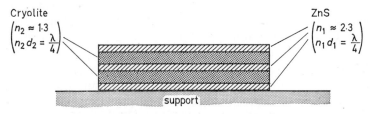

FIG. 7.14. High-reflection dielectric multilayer coating.

Fig. 7.15 compares the relationship of reflective finesse $\mathscr{F}_R$ and maximum étalon transmissivity $\mathscr{T}_M$ for the two cases of silver and Zn–Scryolite coatings. For metal films, there is naturally a continuous curve $\mathscr{T}_M = f(\mathscr{F}_R)$ corresponding to a continuous variation of film thickness but the optical characteristics of a multilayer film structure are, in principle, uniquely defined and hence are represented by a single point on the graph.

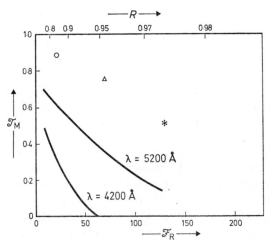

FIG. 7.15. Comparison of values of $\mathscr{T}_M$ and $\mathscr{F}_R$ for a Fabry–Perot étalon having reflecting coatings of silver and of multilayer dielectric films (cryolite/zinc sulphide). The curves are for silver films; the points are for 3-layer (○), 5-layer (△), and 7-layer (∗) dielectric films.

In practice, these characteristics are very dependent on the experimental conditions and the values can be rather widely spread. The points in the diagram represent mean values.

The improvement in performance offered by these multilayer interference films over that of metallic coatings more than compensates for the greater complexity of the deposition process.

In the ultra-violet, the opportunity for dielectric multilayers is more restricted: there is a lack of high-index materials having all the necessary properties. Among those that have been tried are lead chloride, caesium iodide, rubidium iodide and lead fluoride. The last named seems to offer the best prospect: coatings incorporating lead fluoride have shown a reflectivity of 90 per cent with an absorption factor not exceeding 4 per cent in the region of $\lambda = 2500$ Å.[17, 43, 61, 71] At even shorter wavelengths it does not yet appear to be possible to improve on the performance of metallic films, owing to the absence of high-index materials having a sufficiently high transmission, even in thin-film form. A semi-transparent aluminium film with a reflection-enhancing magnesium fluoride film has made it possible to use a Fabry–Perot étalon down to 1800 Å.[4, 5]

Reflecting films for the infra-red do not pose such serious problems: the semiconductors such as silicon, germanium, tellurium and selenium are both highly transparent in their appropriate regions and highly refractive. A triple-layer coating is generally amply sufficient for a high reflection factor; even a single film of high refractive index (tellurium, for instance, with $n \approx 5$) sometimes suffices and has the advantage of comparatively little dependence on wavelength.

Since the high reflectivity of multilayer films is basically due to the effect of interference, it is naturally dependent on wavelength. The reflectivity R reaches a maximum at a predetermined wavelength λ_0, the optical thickness of each layer being adjusted to $\frac{1}{4}\lambda_0$. On either side of λ_0 the reflectivity falls away, more sharply as the number of layers increases. The selective effect is, however, not too serious since the reflectivity remains, in general, sufficiently high over a useful range on either side of λ_0.[42] It is, indeed, possible to achromatize the reflectivity over a wide spectral region by suitably modifying the recipe for the component film thicknesses.[56]

7.21. Some typical Fabry–Perot spectrometers

Instrument for the study of hyperfine line structures

Fig. 7.16 illustrates the principle of an interference spectrometer comprising a Fabry–Perot étalon preceded by a grating monochromator.[1] The principal components are:

Source S
Grating monochromator acting as order sorter
Fabry–Perot étalon with scanning mechanism (illustrated here by a variable-pressure system)
Focusing lens, forming ring fringes in its focal plane
Scanning aperture centred on the focal point of the focusing lens
Detector.

In addition to these elements, there is a number of auxiliary lenses and mirrors, as shown in the diagram.

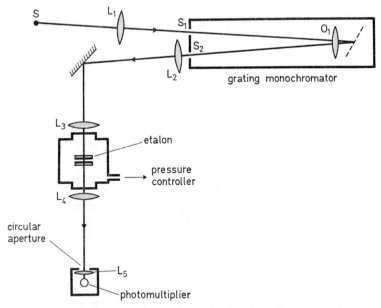

Fig. 7.16. Layout of a Fabry–Perot spectrometer with grating order sorter and pressure scanning (after J. BLAISE, *J. Phys. Radium*, 1958, **19**, 335).

Filtering of orders by means of a single grating monochromator is adequate when the spectrum is not densely populated with lines and when it is unnecessary to eliminate all the parasitic orders of the étalon (see § 7.18). This is normally the case in hyperfine structure work.

The detailed layout of a spectrometer named HYPEAC, built for this purpose in the Aimé Cotton laboratory of the Centre National de la Recherche Scientifique, is shown in Fig. 7.17.[11]

This instrument is intended primarily for resolution studies of emission lines at very high resolution in the region between 2500 Å and 3·5 μm. It was designed so that its high performance could be utilized both by specialists and by relatively untrained operators for routine analysis; the analysis of fine line structure has a strictly practical application in, for example, the isotopic labelling of elements.[2]

The complete mounting is carried on a very rigid framework supported by four anti-vibration springs. The Fabry–Perot étalon is situated in an airtight enclosure so that scanning can be effected by variation of pressure. Temperature is controlled by water circulation.

The light source (A_1B_1 in Fig. 7.17) is a hollow-cathode lamp; the radiation is generated at the cathode spot A_1 and issues from the window at B_1. The grating B_2 with slits A_2 and A_3 and a mirror M_3 constitute an Ebert–Fastie monochromator.

The relay optics are entirely reflective, apart from the lens L_7. This makes the whole system achromatic so that no readjustments are required when changing from one part of the spectral range to another. The cathode spot A_1, the slits A_2 and A_3, and the étalon plates A_4 are optically conjugate; so also are the window B_1 and the grating B_2, the scanning aperture B_4 and the receiving surface of the detector B_5. A supplementary diaphragm aperture B_3 in front of the étalon and conjugate with B_4 eliminates stray light.

The usable angular beam aperture is very often limited by the grating monochromator; in this case the aim is to utilize the available aperture in such a way that the étalon is irradiated at the maximum aperture $\Omega = 2\pi/R$ and that the irradiated area of the plates is as small as possible. It can be shown that to meet these conditions the optical system between M_4 and M_5 should be afocal, with a magnification determined by the specified resolving power $\mathscr{R}$. For this reason mirror M_4 with the two small plane mirrors m_1 and m_2 are mounted on an auxiliary

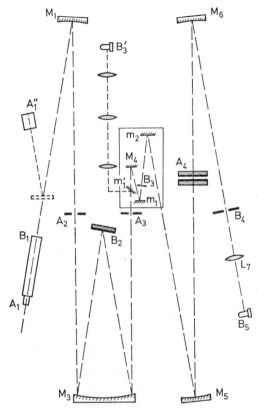

FIG. 7.17. Diagram of HYPEAC Fabry–Perot spectrometer (after R. CHABBAL and P. JACQUINOT[11]).

removable platform to form a unit interchangeable with others of different magnifications.

The Figure also shows an auxiliary source A_1'' and a second detector B_3'. This auxiliary source is used for setting up the instrument. The detector B_3' receives a fraction of the flux emerging from the monochromator via the small plane mirror

178   SPECTROSCOPY AND ITS INSTRUMENTATION

m'_1; its purpose is primarily to monitor the intensity of the source but it can also be used to record the spectrum generated by the monochromator.

Fig. 7.18 is an example of a record obtained with a spectrometer of this type.

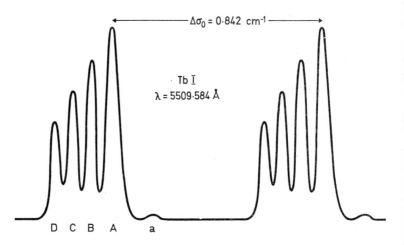

FIG. 7.18. Record of the hyperfine structure of a line in the arc spectrum of terbium. $\Delta\sigma_0$ is the free spectral range.

Single-band Fabry–Perot spectrometer

The examination of complex absorption or emission spectra calls for a spectrometer having a single pass band and a high luminosity; if a Fabry–Perot étalon can be employed without the auxiliary monochromator, full advantage can be taken of its high resolving power and the performance will be adequate. The monochromator can be replaced by a number of étalons in series, that with the largest spacing determining the resolving power and the others eliminating the unwanted orders (see § 7.18).

Fig. 7.19 shows an instrument of this kind[62] using three étalons preceded by an interference filter.† If the spacing of the

† An interference filter is itself a simple Fabry–Perot étalon of very small spacing, the separating layer being a solid transparent film (this filter is sometimes termed a Fabry–Perot filter).

principal étalon E_1 is e_1, those of the other two are $0.8831e_1$ and $0.7244e_1$; these ratios have been chosen to give the maximum suppression of the unwanted orders. With $e_1 = 3$ mm the resolving power is of the order of 350 000.

This instrument has found several applications in astrophysics, particularly in the examination of solar radiation by sodium atoms in the upper atmosphere. As an absorption spectrometer this instrument offers a luminosity 50 to 100 times higher than that which would be obtained at equal resolving power with a mounting of the monochromator/étalon type shown in Fig. 7.16.[69] This advantage naturally diminishes as the demand on resolving power is reduced.

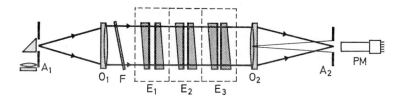

FIG. 7.19. Fabry–Perot spectrometer comprising an interference filter F and three étalons having spacing $e_1 = 3$ mm, $e_2 = 0.8831e_1$ and $e_3 = 0.7244e_1$ (after J. E. MACK, D. P. McNUTT, F. L. ROESLER and R. CHABBAL, *Appl. Optics* 1963, **2**, 873–85).

Fast-scanning spectrometer

Some sources of radiation emit for only a very short time or possess characteristics that change rapidly; plasmas are typical examples of such sources. The spectral analysis of the structure of an emission line must therefore be effected at high speed—one microsecond for the recording of a line profile is not unusual. This rate of scan is feasible with a Fabry–Perot étalon.

A suitable arrangement is shown in Fig. 7.20, in which one plate P_1 of the étalon is cemented to the end of a tube of piezo-electric ceramic (barium titanate).[30] The inner and outer surfaces of the tube are metallized to serve as electrodes; an alternating voltage applied to these electrodes induces longitudinal vibrations of the tube. A collar CC' clamped round the tube at a nodal circumference holds the tube in position. The excitation frequency is chosen to match a resonant frequency of the tube so that the amplitude of vibration may be sufficient for the range of scan required from the étalon. The output from the photomultiplier PM is displayed by a cathode-ray oscilloscope and, with a sufficiently fast time scan on the oscilloscope, a line profile can be traced in 10^{-7}.

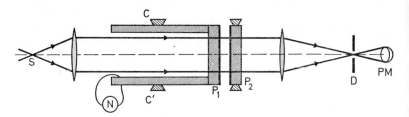

FIG. 7.20. Fast-scanning Fabry–Perot étalon (after J. COOPER and J. R. GREIG[30]).

Multichannel spectrometers

An alternative method of analysing spectral emissions of very short duration is to employ a multi-channel spectrometer. All the spectrometers described so far scan the spectrum in time, whereas the spectrographs display the spectrum in space and the flux representing each spectral element is integrated during the whole time of the exposure. In this respect, the normal spectrometer is at a disadvantage compared with the spectrograph because, in dividing the time T available for a spectral scan by the number of elementary spectral intervals N to give a time T/N for the recording of the energy in a single element, the signal/noise ratio is only divided by $\sqrt{N}$. This leads to greater uncertainty in the measurement of emissions of short life and low energy level.

In order to make available the whole period T for measuring each of the N spectral intervals a spectrometer of the Fabry–Perot type must obviously have N independent detectors in simultaneous action. This presents some nice technical problems but several solutions have nevertheless been shown to be feasible.

For an étalon of finesse $\mathscr{F}$ the free spectral range covers $\mathscr{F}$ elementary intervals. The elementary intervals are represented in the pattern of ring fringes by annular zones, of which there are $\mathscr{F}$ in the annular space between two fringes of neighbouring orders: the problem is to direct the flux in each area on to a separate detector. To this end a system of concentric mirrors, one for each zone, has been used by some workers.[48, 49, 65] These mirrors are inclined, each at its appropriate angle, to reflect the selected flux via a field lens to one of the detectors. One practical mounting in this form used twelve photomultipliers distributed on a circle centred on the axis of the fringe system. In a sixteen-channel spectrometer recently described, a Fresnel lens takes the place of the mirrors.[49]

Another way of focusing the flux from the annular zones is to make use of the peculiar property of the lens form known as the Axicon.[80] This is a plano-conical lens and is mounted behind the focusing lens but within its focal length.[57, 58] Its effect is to introduce a large amount of spherical aberration, rays passing through different zones in its pupil being concentrated at different points along the axis of the lens. Thus when mounted as described above the Axicon focuses the flux in each elementary spectral interval at its particular focal point where the energy can be measured either by interposing a very small detector (such as a photodiode) or by leading it off in a fibre-optic light guide to a photomultiplier.

7.22. Spherical Fabry–Perot étalon

Although the Fabry–Perot étalon spectrometer is, from the point of view of luminosity, greatly superior to the slit spectrometer it is still subject to the rule that resolving power and luminosity vary in inverse ratio, the product $\mathscr{L}\mathscr{R}$ being dependent, for all practical purposes, only on the usable area of the étalon. It follows that if $\mathscr{R}$ is raised to a very high value (for example, several million) the luminosity may fall to an unacceptably low level; this only accentuates the difficulty of working with the low-power sources usually associated with studies involving very high resolving powers.

It is therefore very desirable to escape from the restraint

imposed by the reciprocal relationship of $\mathscr{L}$ and $\mathscr{R}$. Now, the underlying cause of this relationship is the fact that the étendue of the beam passed by the scanning aperture is proportional to $1/\mathscr{R}$, since the angular radius of the aperture varies as $1/\mathscr{R}$. Hence it is the necessity of isolating just one of the interference rings that leads to the restriction on luminosity. This suggests that a possible way of evading this restriction is to eliminate the unwanted rings; they only exist because the path difference in the étalon varies with angle of incidence, a property that is not utilized in the photoelectric form of the Fabry–Perot spectrometer. The ideal interference system for this type of spectrometer is therefore one in which the path difference is independent of the angle of incidence; under these conditions the focal plane of the focusing lens will be uniformly illuminated and the scanning aperture becomes redundant.

The spherical Fabry–Perot étalon invented and put into practice by P. Connes[23, 25] almost completely satisfies the path difference condition. It is an afocal system of spacing e composed of two concave spherical mirrors, the centre of curvature of each being at the vertex of the other (Fig. 7.21). The lower half of each mirror is fully reflecting while the upper half is semi-reflecting. Any incident ray IJ gives rise, according to Gaussian optics, to a directly transmitted ray with which is combined a series of interfering multiple reflections following the path JKLIJ. The path difference JKLIJ is always the same (approximately $4e$) whatever the angle of incidence, so that no fringes are formed and it is theoretically permissible for the incident beam to have an unlimited angular aperture.

This simple analysis is in fact only valid if the Gaussian approximations are acceptable. A rigorous analysis must take into account the third-order aberrations which reintroduce interference fringes and limit the usable étendue: in consequence, two identical circular apertures have to be mounted close to the vertices C_1 and C_2 of the mirrors. The significant attribute of these apertures is that their diameter increases with the étalon spacing e. This being so, it can be shown that the admissible étendue of the beam and correspondingly the limiting luminosity is proportional to e. Thus *the luminosity of a spherical étalon varies in proportion to its resolving power.*

The diameters of apertures suitable for currently used étalon spacings are in the region of a few millimetres, which facilitates the provision of mirrors of very high quality. In this case the limitation on the dimensions of the étalon plates is imposed purely by theory rather than by technical difficulties. The disadvantage of this form of étalon is that since the spacing is a function of the curvature of the mirrors a variation of the resolving power involves a change of mirrors.

The result of exchanging the restraint of a constant $\mathscr{LR}$ value for the constant $\mathscr{L}/\mathscr{R}$ ratio of the spherical étalon is that the plane étalon has the advantage of greater luminosity at medium resolving powers but the spherical étalon becomes greatly

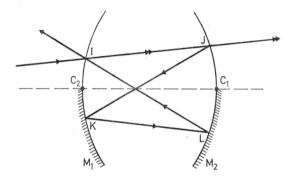

FIG. 7.21. Ray diagram of the Connes spherical Fabry–Perot étalon.

superior in luminosity at high resolving powers. The crossover point is arrived at thus. A plane étalon of spacing e has the same luminosity as a spherical étalon of equal resolving power when the diameter of its plates is $1\cdot 4e$. A reasonable practical limit for this diameter is 70 mm, giving a spacing of 50 mm; this spacing corresponds to an order of interference of about 200 000 at $\lambda = 5000$ Å and a resolving power of a few million. The spherical étalon thus takes over from the plane étalon only at very high resolving powers but then the relative gain in luminosity increases very rapidly, since it is proportional to $\mathscr{R}^2$.

This instrument is therefore only required in spectroscopy for the examination of lines demanding a very high finesse, such as those emitted or absorbed by an atomic beam; then it is invaluable because the radiance of such sources is very weak.[50] Another use of the spherical Fabry–Perot étalon is as a monochromator having a very narrow passband.[27, 28]

FABRY–PEROT ÉTALON SPECTROGRAPH

7.23. *Introduction*

We have so far only discussed Fabry–Perot étalon systems using flux detectors (photomultipliers, photoresistive cells and so on), yet almost from its inception in 1897 this étalon was employed in spectroscopy, during a period when the only practicable detector was the photographic plate. Thus the techniques of Fabry–Perot étalon *spectrography* became highly developed during a long period of study of hyperfine line structures. These methods are still in use and have recently been supplemented by a completely new method combining the advantages of the étalon and of image receptors in a more elegant manner.

7.24. *Classical mounting*

Fig. 7.22 indicates the layout of the classical combination of Fabry–Perot étalon and spectrograph; a normal spectrograph is mounted with its slit S in the focal plane of the focusing lens O_2 and on a diameter of the interference ring system.

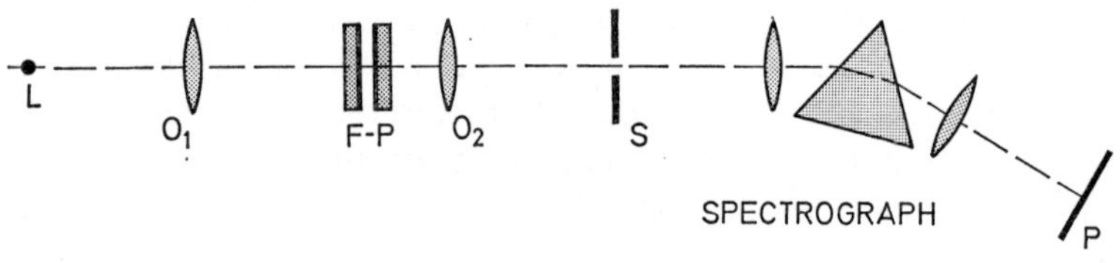

FIG. 7.22. Combination of Fabry–Perot étalon with spectrograph in classical mounting.

Each quasi-monochromatic radiation emitted produces an image on the photographic plate of the section of the corresponding ring fringe system admitted by the entrance slit of the spectrograph; the width of this slit need only be limited to an extent necessary to avoid overlapping of rings due to radiations of neighbouring wavelengths. The radii, intensities and widths of the hyperfine components of the rings are measured on the photograph by means of a microdensitometer.

The luminosity of this system is very much higher than that of the grating spectrograph of identical resolving power, even for only medium resolving powers. In addition this method has

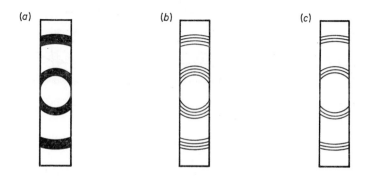

FIG. 7.23. Appearance of spectrum lines photographed with the mounting shown in Fig. 7.22.
(*a*) Broad line; (*b*) 4-component line; (*c*) 3-component line.

the obvious advantage over photo-electric spectrometers that it permits the simultaneous recording of the structure of a large number of lines.

The photographic method does, however, have several disadvantages. The first is that it uses a non-linear detector which also has a poor quantum efficiency; the second is that only a small fraction of the total flux represented by a ring is registered, whereas the photoelectric spectrometer utilizes the whole ring. To overcome this disadvantage a new type of instrument was devised by R. Chabbal, known as the SIMAC. The defects of the photographic plate are not an inherent feature since the plate can be replaced by an electronic image tube.

7.25. The SIMAC

The SIMAC (from the French: Spectrographe Interférentiel Multicanal Aimé Cotton) is designed to accommodate a beam of which the étendue is as large as for the photoelectric spectrometer while preserving the fundamental advantage of the

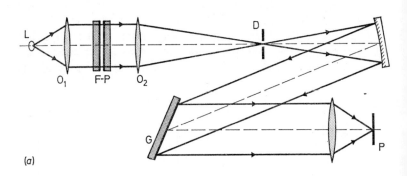

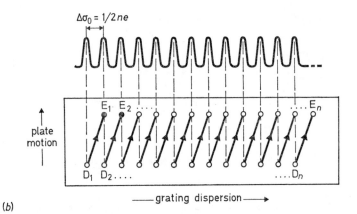

FIG. 7.24(a). Principle of the SIMAC (after R. CHABBAL and R. PELLETIER[12]).
FIG. 7.24(b). Formation of spectrum by the SIMAC.

spectrograph, namely its capacity for recording a large number of spectral elements simultaneously.[12, 13]

In the SIMAC (Fig. 7.24) the central area alone of the fringe system is selected by means of a circular aperture, as in a Fabry–Perot spectrometer. In the present case, however, the flux transmitted by the instrument is passed into a grating spectrograph, the aperture D normally taking the place of the entrance slit (other ways of coupling the étalon and the spectrograph are sometimes used). If the source has a continuous spectrum a series of images $D_1, D_2, \ldots, D_n$ of the aperture D is formed on the photographic plate aligned in the direction of dispersion of the grating. Each image corresponds to an étalon pass band: the spectrograph thus analyses the channel spectra given by the interferometer and the distance between two successive images corresponds to the free spectral range of the interferometer.

To scan the spectrum the spacing of the étalon is varied continuously while the photographic plate is traversed at a constant rate in its own plane in a direction perpendicular to that of the grating dispersion. Each of the images $D_1, D_2, \ldots$ then moves along a straight path $D_1E_1, D_2E_2, \ldots$ and the variations in irradiance (and hence photographic density) along these paths represent the corresponding variations of flux transmitted by the étalon. The scan is complete when the interval between two successive orders has been covered, corresponding to a change of $\frac{1}{2}\lambda$ in the étalon spacing. The information so recorded is the same as that provided by a photo-electric spectrometer but the present method has the considerable advantage that the whole spectrum, however wide and complex, is recorded in a single sweep of the distance between two successive resonant spacings of the étalon. For example, the free spectral range of an étalon of spacing $e = 10$ mm is $\Delta\sigma_0 = 1/2e = 0.5$ cm^{-1}, the étalon having a resolving power of 10^6 at 5000 Å. Its instrumental profile has about 20 000 maxima in the range 3750–6000 Å: in the SIMAC all these 20 000 spectral intervals are scanned simultaneously. The restriction on resolving power imposed by the photographic plate must, of course, be taken into account but this is in fact not a serious limitation because the function of the grating spectrograph is only to separate the interferometer orders. The focal length of the focusing lens can be short (perhaps 200 mm) and the lens may consequently have a high relative aperture (for example $f/2$). The étendues of étalon and spectrograph may thus be matched, a feature of fundamental importance from the point of view of luminosity.

The use of an electronic camera instead of a photographic

plate still further increases the performance of this instrument; in this case, the short focal length of the focusing lens is also an advantage, since a very complex spectrum containing several hundred thousand spectral elements can be concentrated on the receiving surface of the camera. With a detector having this high output, the SIMAC can record a wide and dense spectrum in a very short time and with high resolution, making it particularly valuable for the study of sources having a short lifetime. The accuracy of intensity measurement is, however, not so high as for the photoelectric (photomultiplier) spectrometer because the sensitivity of the electronic camera varies from point to point on its surface. Two examples of SIMAC spectra are shown in Plate 8.

7.26. *The Fabry–Perot étalon in astronomy*

The high luminosity of the Fabry–Perot étalon makes it particularly useful for the spectroscopy of very weak sources so that its value to astrophysics needs no emphasis.

Among the numerous subjects that have been investigated with the help of the SIMAC there is one to which it is particularly well adapted—the study of the emission spectra of nebulae. Certain regions of interstellar space emit radiation in which the H_α line of hydrogen predominates. The areas emitting this radiation may easily be mapped by selective reception of this radiation (using, for instance, a narrow-band interference filter) but a Fabry–Perot étalon makes it possible to measure the wavelength of the line with great precision. Now, this wavelength generally differs slightly from that of the radiation generated in the laboratory with a hydrogen discharge tube; the difference is due to the Doppler effect resulting from the relative motion of the emitting gas. It is thus possible to give precise values to the radial velocities of the different areas of a nebula and hence to gain knowledge of its motion and evolution.

As early as 1911, Fabry, Buisson and Bourget had studied by this means the Orion nebula, which is particularly bright and was the only one capable of being measured by the equipment available at the time. Their technique, considerably improved by G. Courtès, now makes it possible to examine emitting regions of which the magnitudes of radiance and apparent diameter are very low.[31, 32]

The mounting used for these measurements is in principle that shown in Fig. 7.2.5 The combination of field lens L and collimator C forms an image of the telescope mirror on the étalon, the latter being at the entrance pupil of the objective O. Thus all the light collected by the telescope is utilized.

The image of the sky and the interference rings are superimposed in the focal plane P of the objective O, where they are recorded by a photographic plate. It is thus possible to determine the exact wavelength of the emission at different areas of the nebula by measuring the diameters of the rings relative to those of reference rings produced by the radiation from a hydrogen discharge lamp.

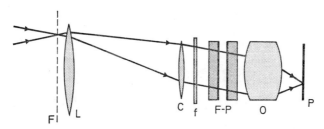

FIG. 7.25. Mounting for the study of interstellar radiation by means of a Fabry–Perot étalon.
F: focal plane of telescope mirror.   C: collimator.
L: field lens.   f: filter.
F–P: Fabry–Perot étalon.   O: camera objective.
P: plate.

In the mounting shown in Fig. 7.25 each image point in the focal plane F produces at the étalon a parallel beam which covers the whole surface of the étalon plates. The étalon defects, including particularly those of the multidielectric coatings, are therefore incapable of introducing any systematic error into the measurements.

The focal length of the objective O is chosen to be of such a value that the width of a ring is approximately equal to the resolution limit g of the emulsion, so ensuring the best compromise of luminosity and resolving power (see § 2.9). A typical focal length is 25 mm.

Plate 9 is an example of a photograph taken with this instrument. Its subject is a nebula with a relatively high concentration of emissive hydrogen (Messier 33). The radiation selected is H_α and the telescope is the 1·93 m reflector at the Observatoire de Haute-Provence. This photograph was the first to show the distribution of the H_α emission line over the nebula. One of the spiral arms is clearly seen in the top left-hand quadrant of the picture.

The distribution of velocities in this nebula has been measured by means of the Doppler effect at more than 1000 points and thus all its dynamic properties have been recorded, including its period of rotation (10^7 years) and the discontinuity in angular velocity in the regions forming the spiral arms.

Variations of the mounting described, employing very short-focus objectives, have been specially developed for the study of nebulae of very small apparent diameters.[33]

INTERFERENCE SPECTROSCOPY BY THE 'SISAM' METHOD

7.27. *Introduction*

For a given resolving power, a Fabry–Perot étalon has a much greater étendue than a slit spectrometer and consequently has a much higher luminosity. On the other hand the interferometer is at a disadvantage because of its very restricted spectral range, which enforces the addition of an auxiliary monochromator.

In the grille spectrometer we have already had an example of a dispersing instrument which succeeds in retaining the large free spectral range of a grating while transmitting much more flux than the corresponding slit spectrometer. In point of fact the first instrument designed to meet this specification was the SISAM† interferometer, due to P. Connes, in which wavelength sorting is achieved by selective modulation.

† SISAM: Spectromètre Interférentiel à Sélection par l'Amplitude de Modulation.

7.28. Principle of the SISAM interferometer[24,26]

This instrument is a Michelson interferometer in which the two mirrors are replaced by two identical gratings G_1 and G_2.† The rulings of these gratings are perpendicular to the dispersion plane (the plane of Fig. 7.26) and are so oriented that their dispersions are equal and in the same sense.

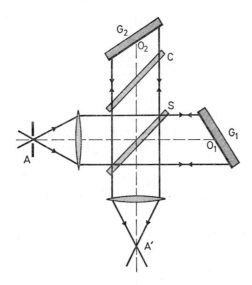

FIG. 7.26. Layout of SISAM interferometer.

Suppose the instrument receives a monochromatic parallel beam of wavenumber σ from a source aperture A. When σ has the particular values σ' given by $\sigma' = k/2c\sin i$, the diffracted beam is parallel to the incident beam and then the instrument has exactly the same properties as a normal Michelson interferometer adjusted to give interference rings at infinity. σ' will henceforward be called the setting wavenumber.‡

---

† It is also possible to use reflecting prisms or combinations of prisms and gratings. Although these alternatives may have an advantage in luminosity, they are rarely used and will not be considered here.

‡ The order k can of course have several values, but in practice it is easy to separate the spectra of differing orders. We shall assume that this is done so that σ' may be given a unique value.

For any radiation for which σ differs from σ' the interferometer is not in adjustment, in the sense that the wavefront diffracted by each grating makes an angle ε with the incident wavefront: the two diffracted wavefronts are then mutually inclined at an angle 2ε in the image space. The expression for the emergent flux is then†

$$\Phi = \tfrac{1}{2}\Phi_0\{1 + P\cos(2\pi\sigma\delta)\} \qquad (7.13)$$

Φ_0 being the incident flux, δ the path difference and P having the value

$$P = \frac{\sin(2\pi\sigma\varepsilon a)}{2\pi\sigma\varepsilon a} \qquad (7.14)$$

(a is the width of the beams reflected by G_1 and G_2). The flux Φ is thus seen to vary sinusoidally with the path difference δ, the corresponding depth of modulation having a maximum value of unity when $\varepsilon = 0$ and decreasing rapidly when this angle increases.

To make use of the fact that Φ depends on δ it is only necessary to cause the optical path of one interfering beam to vary as a linear function of time.

This may be done in a variety of ways but the simplest is to rotate the compensator plate. Then, if $\delta = vt$, the emergent flux is given by

$$\Phi = \tfrac{1}{2}\Phi_0\{1 + P\cos(2\pi\sigma vt)\} \qquad (7.15)$$

This flux is therefore modulated at a frequency vt but since P is zero for all values of σ excepting those very close to the setting wavenumber σ' the modulation is only detectable in radiations of wavenumber contained in a narrow band about σ'. The interferometer is thus a selective modulator for this radiation, whence its name.

Now consider the source of radiation to have a complex spectrum. The emergent flux is then modulated only in the narrow band centred on the setting wavenumber and is unmodulated in the rest of the spectrum. There is no problem in isolating the modulated component of the output signal and

---

† It is assumed that flux losses in the instrument are negligible and that the transmission and reflection factors of the beam-splitter are both 0·5.

INTERFERENCE SPECTROSCOPY

measuring its amplitude. If the modulated signal were of zero amplitude for all radiation other than that having precisely the setting wavenumber σ' the measured magnitude would be exactly proportional to the monochromatic flux $\Phi_\sigma(\sigma')$. In practice this is of course not quite true as the SISAM has, like all other spectrometers, an instrumental profile of finite width.

The spectrum is scanned by simultaneously rotating both gratings about their axes (O_1, O_2) in the same sense and at the same angular velocity; the rotation of the compensator plate then modulates each spectral element in turn and the spectrum is recorded in the normal way.

7.29. *Theoretical instrumental profile*

We assume the interferometer to be receiving a parallel beam of monochromatic radiation. The recorded signal is proportional to the depth of modulation P of the emergent flux, so it is the variation of this quantity as a function of σ that gives the instrumental profile.

We already have

$$P = \frac{\sin(2\pi\sigma\varepsilon a)}{2\pi\sigma\varepsilon a}$$

By definition, $\varepsilon = 0$ when $\sigma = \sigma'$. For a different value of σ it is clear that $\varepsilon = (\sigma - \sigma')\, d\theta/d\sigma$ if $d\theta/d\sigma$ is the angular dispersion of the gratings. We also know the theoretical resolving power of the gratings to be

$$\mathscr{R}_0 = \sigma/\Delta\sigma = a\sigma^2\, d\theta/d\sigma$$

It follows that

$$P = \frac{\sin\{2\pi(\sigma - \sigma')/\Delta\sigma\}}{2\pi(\sigma - \sigma')/\Delta\sigma} \qquad (7.16)$$

$\Delta\sigma$ representing the elementary spectral interval that corresponds to the theoretical resolving power $\mathscr{R}_0$ of the SISAM gratings.

This function P has a familiar shape (see § 4.2, Fig. 4.4); its width measured on the wavenumber axis is equal to $\Delta\sigma$.

7.30. *Effective resolving power and luminosity*

We have so far assumed the instrument to be receiving a parallel beam. In the practical case the aperture A in the focal plane of the objective must be large enough to pass a useful amount of flux. It can be shown that the optimum aperture is circular, at least to a first approximation, with an angular radius $\alpha = \sqrt{(2/\mathscr{R}_0)}$, just as in the case of the Fabry–Perot étalon. The practical instrumental profile is therefore the convolution of the theoretical profile equation (7.16) and a rectangular function of the same width representing the influence of the aperture. The effective resolving power of the SISAM is practically $\mathscr{R} = \mathscr{R}_0$, so that it is equal to that of one of the gratings.

The optimum value of the angular radius α of the aperture is thus determined by the value of $\mathscr{R}_0$: this means that in contradistinction to slit spectrometers the SISAM has a constant resolving power. It always operates at a resolving power close to the theoretical resolving power of the gratings. If the aperture is opened up to a value above the optimum there is little gain in luminosity to set against the significant loss of resolving power.

The solid angle $\Omega = 2\pi/\mathscr{R}_0$ subtended by the SISAM aperture at the collimator is the same as that for a Fabry–Perot étalon of the same resolving power; the luminosity is therefore much greater than that of the corresponding slit spectrometer.

The SISAM is particularly well suited to act as the monochromator for a Fabry–Perot spectrometer.[45]

7.31. *Apodization of the instrumental profile*

The theoretical instrumental profile given by equation (7.16) possesses a serious disadvantage, namely, the presence of secondary maxima of considerable magnitude. It is, therefore, in theory necessary to apodize† the profile, that is, to reduce the

---

† A term introduced by P. JACQUINOT (see *Proc. Phys. Soc.*, 1950, **B63**, 969).

magnitude of unwanted maxima. This is in general achieved by suitably modifying the distribution of the radiation amplitude in the pupil of the optical system. In the present case, however, the aperture itself introduces a sufficient degree of apodization.

7.32. *Region of application of the SISAM*

The SISAM photodetector receives not only the signal flux within the resolved spectral interval but also a considerable amount of unmodulated parasitic radiation from neighbouring regions of the spectrum. In this respect it is in the same case as the detector in the grille spectrometer discussed in § 6.21 and the conclusions reached there are equally true here: in order that the parasitic flux may not adversely affect the signal/noise ratio the source of noise must mainly be restricted to the detector and must not be associated with the radiation. This means that the SISAM is most advantageously employed in the infra-red.

7.33. *Practical details of construction*[26, 44, 63, 73]

Modulation by interference calls for a variation of path difference which is linear with time. This is most easily achieved by rotating the compensating plate at a constant rate; the angular displacement required is, however, very small and consequently the direction of rotation must be alternated. The result is a symmetrical saw-tooth pattern of variation in path difference.

The fact that the modulation frequency is proportional to wavenumber is turned to advantage as a means of separating the various orders of the grating spectra: it is easy to accept only the signal due to the desired order by means of an electrical band-pass filter tuned to the appropriate frequency, thus obviating the use of an auxiliary monochromator.

The inclusion of a second diaphragm aperture A' in the focal plane of the focusing lens to coincide with the image of aperture A is theoretically unnecessary but in practice has the desirable effect of eliminating a large part of the unmodulated parasitic flux.

FOURIER SPECTROSCOPY

7.34. *Introduction*

The fundamental distinction between dispersing instruments having an image receiver (spectrographs) and those employing a flux detector (spectrometers) has already been drawn (see § 7.21). In the first case the elements of the spectrum are dispersed in space and all are received simultaneously whereas in spectrometers the elements of the spectrum are recorded sequentially, only one element being received at one moment of time. From this point of view the energy emitted by the source is more efficiently utilized in spectrographs than in spectrometers, and there is a gain in signal/noise ratio of $\sqrt{N}$ when N spectral elements are measured simultaneously instead of sequentially.[54]

This gain is only realizable when the measurements of the N elements are independent of each other, so that the signal/noise ratio ascribable to one spectral element is not affected by the presence of neighbouring elements. This condition is obviously met when the elements are measured with separate detectors or with independent areas of the same detector, as in spectrographs. Multiple detectors are, however, only feasible when the number of elements is quite small: it is so, for instance, in the (typically) ten-channel Fabry–Perot spectrometer described in § 7.21. An extension of this method to the study of an extended spectrum containing several thousand elements is quite inconceivable; some method had therefore to be found which made possible the simultaneous recording of the flux in a large number of spectral elements with a single detector. The solution is found in an application of the principle of multiplexing commonly applied in telecommunication practice, which allows a number of communication channels to be carried by a single line.†

To label the signals corresponding to N separate spectral radiations it is only necessary to modulate each radiation in a sinusoidal mode at a different frequency before it reaches the

---

† Multiplex spectrometers are said to possess the Fellgett advantage from the name of one of the authors who explicitly drew attention to the advantage offered by optical multiplexing.

INTERFERENCE SPECTROSCOPY

detector. The output signal is thus composed of N superimposed sinusoidal signals, each having an amplitude proportional to the flux representing the intensity of its own spectral element. The spectrum of the source is subsequently derived by carrying out a Fourier analysis of the output signal. This method thus acquires the designation of Fourier spectrometry.

7.35. Modulation by two-beam interferometer

A two-beam interferometer (normally a Michelson) is particularly well suited for modulating the spectral radiations; this may be shown by supposing a monochromatic radiation of wavenumber σ to be received by the interferometer. If this incident flux is Φ_0, the emergent flux† is

$$\Phi_1 = \Phi_0 \cos^2(\pi\sigma\delta) = \tfrac{1}{2}\Phi_0\{1+\cos(2\pi\sigma\delta)\}$$

where δ is the path difference.

If δ varies linearly with time at a rate v, the flux becomes

$$\Phi_1 = \tfrac{1}{2}\Phi_0\{1+\cos(2\pi\sigma v t)\}$$

The output flux is therefore the sum of a constant term and of a term varying sinusoidally in time at a frequency $\nu = \sigma v$ proportional to the frequency of the radiation.‡ The required result is therefore achieved.

7.36. Principle of Fourier spectrometry[39, 52]

We now allow the interferometer to accept heterochromatic radiation; $\Phi_\sigma(\sigma)$ will designate the monochromatic incident flux corresponding to a wavenumber σ. The flux transmitted within an elemental spectral interval $d\sigma$ is $\Phi_\sigma\, d\sigma$ and the total emergent flux is:

$$\Phi_1(\delta) = \frac{1}{2}\int_0^\infty \Phi_0(\sigma)\{1+\cos(2\pi\sigma\delta)\}\, d\sigma \qquad (7.17)$$

† The instrument is assumed to satisfy the conditions specified in the footnote on p. 192.
‡ A similar result has already been noted for the SISAM but in that case the frequency of modulation was an arbitrary quantity whereas here it is of fundamental significance.

The variable part of this expression

$$\Phi(\delta) = \frac{1}{2}\int_0^\infty \Phi_0(\sigma)\cos(2\pi\sigma\delta)\,d\sigma \qquad (7.18)$$

is, apart from the coefficient $\frac{1}{2}$, simply the cosine Fourier transform of the function $\Phi_\sigma(\sigma)$, itself proportional to the source function $L_\sigma(\sigma)$ which represents the variation of source radiance as a function of wavenumber.

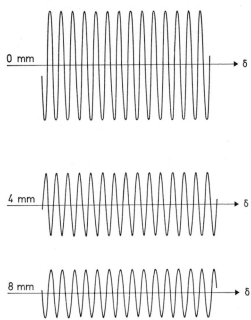

(a)

FIG. 7.27. Examples of interferograms, showing the influence of spectral bandwidth.
(a) Radiation of wavelength $\lambda = 5461$ Å emitted by a low-pressure mercury vapour lamp. The upper curve corresponds to a path difference δ close to zero, the middle curve to $\delta \approx 4$ mm and the lower curve to $\delta \approx 8$ mm.

We can say, therefore, that if in a two-beam interferometer receiving a heterogeneous radiation the path difference δ is varied and the variable part of the receiver output is recorded, the resulting *interferogram* is proportional to the cosine Fourier transform of the function $L_\sigma(\sigma)$.

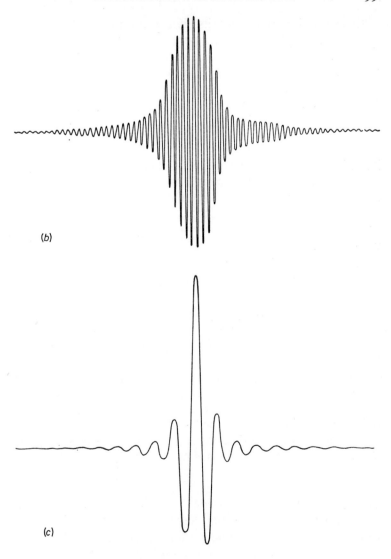

FIG. 7.27—*continued*.
(b) Spectral bandwidth $\Delta\lambda \approx 1000$ Å at $\lambda \approx 10\,000$ Å.
(c) Radiation from an incandescent lamp, the detector being a lead sulphide cell (after J. CONNES, *Rev. Optique Théor. Instrum.* 1961, **40**, 49–51).

Since Fourier transformation is a reciprocal operation, the problem of analysing the incident radiation is, in principle, solved: to obtain the source function $L_\sigma(\sigma)$ it is only necessary to calculate the Fourier transform of the interferogram.

As the recorded function (7.18) is, apart from a coefficient, identical with the function

$$\mathscr{I}(\delta) = \int_0^\infty L_\sigma(\sigma)\cos(2\pi\sigma\delta)\,d\sigma \qquad (7.19)$$

it is this latter expression which will, from now on, be designated the interferogram. $\mathscr{I}(\delta)$ is clearly an even function while $L_\sigma(\sigma)$ is defined only for positive wavenumbers since negative values of σ have no physical significance.

The inversion of the Fourier integral (7.19) is, for positive values of σ:

$$L_\sigma(\sigma) = 2\int_{-\infty}^{+\infty} \mathscr{I}(\delta)\cos(2\pi\sigma\delta)\,d\delta$$

$$= 4\int_0^\infty \mathscr{I}(\delta)\cos(2\pi\sigma\delta)\,d\delta \qquad (7.20)$$

The spectrum is therefore obtained, as anticipated, by calculating the Fourier transform of the interferogram.

Unfortunately the extraction of the source function by this operation is often extremely tedious.

The shape of the interferogram depends, of course, on that of the source function. In the simple case of a strictly monochromatic incident radiation the function $\mathscr{I}(\delta)$ is sinusoidal. When the incident radiation is confined within a narrow spectral band $\Delta\sigma$, the interferogram takes the form of an attenuated pseudo-sinusoid curve of which the amplitude decreases (with δ) more rapidly as $\Delta\sigma$ increases.

When the incident radiation has a wide spectrum, the curves resulting from the constituent monochromatic radiations are out of phase with each other to varying degrees except for $\delta = 0$, when they are all in phase. The interferogram is then reduced to a few fringes in the neighbourhood of the central maximum at $\delta = 0$.

The cases cited represent the extremes; in general, the amplitude of modulation in the interferogram decreases more rapidly as the spectral range being sampled increases.[18] There is a natural correlation between this and the well-known fact that the decrease in visibility of quasi-monochromatic two-beam fringes as the path difference is increased occurs more rapidly as the bandwidth of the radiation is widened.

In certain cases an examination of the visibility of fringes can give information on the spectral profile of a line without the necessity of actually calculating the Fourier transform of the interferogram. Michelson originally used this method and it has been applied by J. Terrien to the study of spectral lines utilized in metrology; this work will be described later (see § 7.43).

7.37. Theoretical instrumental profile: apodization: resolving power[18]

Equation (7.20) shows that for the calculation of $L_\sigma(\sigma)$, it is necessary to know the function $\mathscr{I}(\delta)$ for all path differences between 0 and ∞. In practice, δ can only be increased to some maximum value δ_M; this means that, instead of having the function $\mathscr{I}(\delta)$, only its product with the rectangular function of width $2\delta_M$ and unit height can be known (since $\mathscr{I}(\delta)$ is an even function it is in fact known between the limits $-\delta_M$ and $+\delta_M$). When the spectrum is calculated the result is, therefore, the Fourier transform $F(\sigma)$ of this product instead of that of $\mathscr{I}(\delta)$. Now, the Fourier transform of the product of two functions is equal to the convolution of their transforms. Hence, instead of the source function $L_\sigma(\sigma)$ we obtain from the instrument output

$$F(\sigma) = L_\sigma(\sigma) * A_0(\sigma) \qquad (7.21)$$

where $A_0(\sigma)$ is the Fourier transform of $\text{rect}(\delta/2\delta_M)$. This equation shows that $A_0(\sigma)$ corresponds to an instrumental profile for the Fourier transform spectrometer. Its familiar formula is

$$A_0(\sigma) = \frac{\sin(2\pi\sigma\delta_M)}{2\pi\sigma\delta_M} \qquad (7.22)$$

This can be taken only as a theoretical instrumental profile since it only takes account of the finite range of path difference. The corresponding theoretical resolving power is $2\sigma\delta_M$, and is thus proportional to the maximum path difference attained in the instrument.

This instrumental profile is the same as that of the SISAM and we have already seen that it has an undesirable shape because of the presence of comparatively large secondary maxima. It is therefore again necessary to apodize the profile. This operation requires the choice of a suitable function $H(\delta)$ to replace the function $\text{rect}(\delta/2\delta_M)$ as a multiplying factor for the interferogram $\mathscr{I}(\delta)$ between $-\delta_M$ and $+\delta_M$. (This amounts to choosing a favourable instrumental profile since this is given by the Fourier transform of $H(\delta)$.) If $H = \text{rect}(\delta/2\delta_M)$ is replaced by a triangular function of unit height and of width $2\delta_M$, so that $H = \text{tri}(\delta/\delta_M)$,† the instrumental profile becomes:

$$A_0(\sigma) = \left\{\frac{\sin(\pi\sigma\delta_M)}{\pi\sigma\delta_M}\right\}^2 \tag{7.23}$$

This profile is a considerable improvement on the original.

Physically this apodization was carried out on the SISAM by placing suitably shaped apertures on the gratings. In the present case apodization can be achieved without loss of radiation by attenuating the signal from the detector to an extent that is variable with the path difference. In comparison with the SISAM system of attenuating the radiation, this method has the advantage that the signal/noise ratio is not changed since the detector noise is reduced with the signal. There is, however, an even simpler method which consists in multiplying the ordinates of the interferogram by suitable factors; this process, sometimes described as mathematical apodization, is carried out by the computer that calculates the spectrum and occupies only a small fraction of the total computation time.

With an instrumental profile of the form given by equation (7.23), having a width $\Delta\sigma = 1/\delta_M$, the theoretical resolving power becomes $\mathscr{R}_0 = \sigma\delta_M$; the apodization has thus had the usual effect of reducing the resolving power.

† The notation $\text{tri}(x)$ represents a triangular function of unit height and of which the base on the x-axis extends from -1 to $+1$. This function is therefore defined as follows:

$$\text{tri}(x)\begin{cases} = 0 & \text{for } |x| > 1 \\ = 1-|x| & \text{for } |x| < 1 \end{cases}$$

Since one interference fringe passes a point in the field each time the path difference changes by λ, $\mathcal{R}_0$ can be evaluated as the total number of fringes scanned when the path difference is changed from 0 to δ_M.

Plate 10 shows the influence of the maximum value of the path difference on resolving power.[74]

In practice a large variation of path difference in a Michelson interferometer can only be obtained by displacing one of the mirrors. Since the angular setting of this mirror must remain strictly unchanged during this movement, high resolving powers demand very high mechanical precision for this movement.

7.38. *Étendue of accepted beam*

So far no account has been taken of the fact that the angular spread of the incident beam is not zero. The acceptable value of étendue depends on the type of interferometer and is therefore not a characteristic peculiar to Fourier spectrometry itself.

In interferometry based on the Michelson principle the angular extent of the beam is limited by a circular aperture of angular radius α placed in the focal plane of the focusing lens (the same device is also used with the Fabry–Perot étalon and the SISAM). If S stands for the effective area of the focusing lens, the étendue of the accepted beam is $U = S\pi\alpha^2$. We now have to find the value of α that gives the optimum combination of resolving power and luminosity; to do this we need an expression for the partial instrumental profile $A_D(\sigma)$ which includes the aperture dimension.

A Michelson interferometer is normally adjusted so that one mirror is parallel to the image of the other formed by the beamsplitter. In this case the fringes are at infinity so that if the path difference for a monochromatic radiation (wavenumber σ_0) is δ_0 at normal incidence, then for rays incident at angle i the path difference is $\delta = \delta_0 \cos i \approx \delta_0(1 - \tfrac{1}{2}i^2)$. The transmission factor is

$$\mathcal{T} = \tfrac{1}{2}\{1 + \cos(2\pi\sigma_0\delta)\} \qquad (7.24)$$

and so this also depends on the angle of incidence. Rays

incident at angles between i and $i+di$ form a beam of solid angle $d\Omega = 2\pi i\, di$. From equation (7.24) the component of flux within $d\Omega$ varying with δ is

$$d\Phi = \tfrac{1}{2}LS\cos(2\pi\sigma_0\delta)\, d\Omega \qquad (7.25)$$

where L is the monochromatic radiance. We have

$$\delta = \delta_0(1 - \tfrac{1}{2}i^2)$$

so that

$$d\delta = -\delta_0 i\, di = -\delta_0\, d\Omega/2\pi$$

Thus

$$d\Phi = -\frac{\pi LS}{\delta_0}\cos(2\pi\sigma_0\delta)\, d\delta$$

so expressing the distribution of flux in terms of path difference rather than of angle of incidence.

The total flux transmitted by an interferometer fitted with a circular aperture of angular radius α is

$$\Phi = \frac{\pi LS}{\delta_0}\int_{\delta_0(1-\frac{1}{2}\alpha^2)}^{\delta_0}\cos(2\pi\sigma_0\delta)\, d\delta$$

$$= \frac{\pi LS}{\pi\sigma_0\delta_0}\sin(\tfrac{1}{2}\pi\sigma_0\delta_0\alpha^2)\cos\{2\pi\sigma_0\delta_0(1-\tfrac{1}{4}\alpha^2)\} \qquad (7.26)$$

which may also be written

$$\Phi = \frac{LU}{2}\cdot\frac{\sin(\tfrac{1}{2}\pi\sigma_0\delta_0\alpha^2)}{\tfrac{1}{2}\pi\sigma_0\delta_0\alpha^2}\cos\{2\pi\sigma_0\delta_0(1-\tfrac{1}{4}\alpha^2)\}$$

For an infinitely small aperture the interferogram would be reduced to the function $\cos(2\pi\sigma_0\delta_0)$. The finite size of the aperture therefore has two effects: a shifting of the phase of the original function by an amount proportional to α^2 and a multiplication of that function by the factor

$$\frac{\sin(\tfrac{1}{2}\pi\sigma_0\delta_0\alpha^2)}{\tfrac{1}{2}\pi\sigma_0\delta_0\alpha^2}$$

It is easy to show that the phase shift results in a translation by an amount $\frac{1}{4}\sigma_0\alpha^2$ along the wavenumber axis. The corresponding correction is thus very simply calculated.

On the other hand multiplication of the interferogram by a factor depending on δ implies the existence of an instrumental profile which will be the Fourier transform of that factor.

The instrumental profile A_D due to the aperture is therefore the Fourier transform of

$$\frac{\sin(\frac{1}{2}\pi\sigma_0\delta_0\alpha^2)}{\frac{1}{2}\pi\sigma_0\delta_0\alpha^2}$$

or, omitting the coefficient,

$$A_D(\sigma) = \text{rect}\left(\frac{\sigma}{\frac{1}{2}\sigma_0\alpha^2}\right) \tag{7.27}$$

This expression is identical with that for the Fabry–Perot interferometer. The width of the rectangular function is $\frac{1}{2}\sigma_0\alpha^2$.

The conclusions in § 7.13 apply equally in the present case. The best compromise between resolving power and luminosity is reached when the two instrumental profiles A_0 and A_D have the same width, that is, when $\frac{1}{2}\sigma_0\alpha^2 = 1/\delta_M = \sigma_0/\mathscr{R}_0$, $\mathscr{R}_0$ being the theoretical resolving power as defined above.

The angular radius of the scanning aperture is now $\alpha = \sqrt{(2/\mathscr{R}_0)}$ as for the Fabry–Perot and SISAM spectrometers. The Fourier spectrometer therefore accepts a beam of the same angular aperture as that in the other two and hence, like them, has the Jacquinot advantage (§ 7.17).

7.39. *Computation of the spectrum*

The spectrum is obtained from the interferogram by a Fourier transformation defined by equation (7.20). The amount of calculation involved in all but the simplest cases makes the use of a digital computer an absolute necessity. At very low resolving powers it is indeed possible simply to record the detector output on magnetic tape and to carry out a harmonic analysis of the tape output by means of a standard frequency analyser. Analogue computers designed specially for the spectrometer have also been proposed but again only for very limited resolving powers.[38] In all other cases digital computation is essential; it is usually carried out on general-purpose machines.

Principle of the method[18]

The interferogram $\mathscr{I}(\delta)$ will first have been multiplied by the apodizing function $H(\delta)$, so to derive the spectrum the integral to be calculated is

$$F(\sigma) = \int_0^\infty \mathscr{I}(\delta)H(\delta)\cos(2\pi\sigma\delta)\,d\delta \qquad (7.28)$$

This integral has to be replaced by a summation based on a number of discrete values taken from the interferogram. In principle this does not affect the accuracy to which the spectrum is known. The spectrum never in fact extends beyond a finite range $\sigma_2 - \sigma_1$ on the wavenumber scale since it is limited either by the emission range itself or by the transmissivity of materials in the instrument or by detector sensitivity. Now, according to the *sampling theorem* all the information contained in a function of which the spectrum has a limited range l is carried by an infinite number of discrete values separated by $1/l$. In the present case, then, all the information contained in the interferogram $\mathscr{I}(\delta)$ can be obtained by taking a number of discrete values of this function at a maximum spacing of $1/(\sigma_2 - \sigma_1)$.

The procedure is therefore to read off values $\mathscr{I}_0, \mathscr{I}_1, \ldots, \mathscr{I}_n$ corresponding to equidistant abscissae $0, h, \ldots, nh$ (where $nh = \delta_M$) and to calculate the sum

$$F'(\sigma) = h(\tfrac{1}{2}H_0\mathscr{I}_0 + H_1\mathscr{I}_1 + \cdots + H_n\mathscr{I}_n) \qquad (7.29)$$

$H_0, H_1, \ldots, H_n$ being the values of the apodizing function at the corresponding values of δ.

When the integral (7.28) is replaced by the sum $F'(\sigma)$, the interferogram can be regarded as having been multiplied by the sum of an infinity of Dirac delta functions having equidistant abscissae $\delta = kh$ forming what is sometimes called a Dirac comb.

$$\rho(\delta) = \sum_{k=-\infty}^{+\infty} \delta(\delta - kh)$$

Now, we know that multiplying an interferogram by a factor that is a function of the path difference amounts to convoluting the instrumental profile of the spectrometer with the Fourier transform of that factor. The Fourier transform of a Dirac comb of spacing h is a comb of spacing $1/h$. The convolution of the instrumental function $A(\sigma)$ with this distribution results in a succession of $A(\sigma)$ functions at $1/h$ spacings.

The effect of replacing the Fourier integral (7.28) by a summation process based on discrete values of $\mathscr{I}(\delta)$ thus gives rise to a modification of the instrumental profile; this now has an infinite number of peaks instead of an isolated maximum. The profile is analogous to that of a Fabry–Perot étalon, the interval $\Delta\sigma_0 = 1/h$ in the case of the Fourier spectrometer corresponding to the free spectral range of the étalon. In consequence, a series of repetitions of the spectrum is generated in the process of recovery of the spectrum.

The situation is, however, somewhat different in the present case from that of the Fabry–Perot étalon in that we are free to choose h.

Sampling interval

The values of h must be small enough to avoid overlap of the reconstructed spectra. The maximum usable sample spacing lies between $\frac{1}{2}(\sigma_2 - \sigma_1)$ and $\frac{1}{4}(\sigma_2 - \sigma_1)$ according to the values of σ_1 and σ_2.† In practice, smaller values of the spacing are chosen from consideration of the effect of background noise on accuracy.[18,19]

Number of sampling points on the interferogram

The interferograms cover the range 0 to δ_M on the path-difference axis, the maximum value being related to a resolved

---

† The period $1/h$ of the instrumental profile might be expected to be $\sigma_2 - \sigma_1$; this is not so because before sampling has introduced the comb function the profile has, not one maximum at σ_0, but two maxima at $+\sigma_0$ and $-\sigma_0$. This arises from the fact that the Fourier transform of $\cos(2\pi\sigma_0\delta)$ consists of two delta functions $\frac{1}{2}\delta(\sigma \pm \sigma_0)$. In the process of reconstruction it becomes necessary to take account of negative values of σ. The calculation produces not only the real spectrum $L_\sigma(\sigma)$ for $\sigma > 0$ but also $L_\sigma(-\sigma)$ for $\sigma < 0$. This is the origin of the values indicated above for the sampling period.

spectral element $\Delta\sigma$ by $\Delta\sigma = 1/\delta_M$. The minimum number of measurements to be taken from the interferogram is δ_M/h and hence varies between $2(\sigma_2-\sigma_1)/\Delta\sigma$ and $4(\sigma_2-\sigma_1)/\Delta\sigma$. These limits may be written $2N$ and $4N$, N being the number of resolved spectral elements contained in the spectrum measured. In practice, as explained above, the number is higher: for example, $10N$.

Time required for computation

For a digital method, the computation time is proportional to the product of the number of points taken on the interferograms and the number N of spectral elements included. The time taken is therefore proportional to N^2 and so increases rapidly with both spectral range and resolving power.

The time factor is a serious handicap for Fourier spectroscopy, especially since in principle it increases its advantage over other methods as the number N of spectral elements sampled is increased. Even with very fast computers the calculation time rapidly becomes embarrassingly long for large values of N. Although special machines have been developed to deal with comparatively low N values,[40,66,74] it seems that the problem of shortening the time of operation on normal computers still has to be solved if large N values are to be used.[21]

7.40. *Practical details of design*

Fourier spectrometers are almost always based on a Michelson interferometer modified to meet the particular requirements of the method. In particular, the path difference must be capable of being varied without disturbing the alignment of the interferometer. This is achieved by replacing the mirrors by reflecting systems which do not give rise to deflections of the reflected beam if they rotate. Suitable reflectors are the corner cube (three plane mirrors mutually at right-angles as at the corner of a cube) or the cat's-eye (a convex and a concave mirror having radii of curvature in the ratio 2:1 and mounted with coincident centres of curvature).[34,55] With such reflectors, displacements of several hundred millimetres can be attained without appreciable disturbance of the fringe alignment.

The interferometer must also be perfectly achromatic so that the path difference does not vary with wavelength. If this condition is not met the path difference is not zero for all wavelengths simultaneously and the result is an asymmetry in the interferogram. The effect of even a very small degree of asymmetry is to modify the instrumental function. Strict achromatism can only be achieved with a perfectly symmetrical interferometer but this is not possible in the Michelson. The reason is that the beam-splitter S and the compensator plate C never have exactly the same thickness; a path difference (which is wavelength-dependent) is thus introduced because one beam is transmitted twice by C and once by S whereas the opposite is the case with the other beam. Moreover, one beam is externally reflected and the other internally reflected at the coated surface of the beam-splitter: the two modes of reflection induce phase shifts that vary differently with wavelength. This again introduces a source of asymmetry.

Perfectly symmetrical interferometers have been described. Fig. 7.28 shows an example of a system which avoids the use of

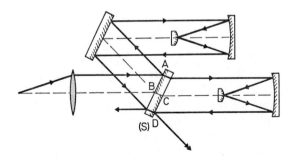

FIG. 7.28. Michelson interferometer modified to symmetrical form incorporating cat's-eye afocal reflectors (after P. JACQUINOT[55]).

a compensator plate by having one half AB of one face of the beam-splitter (S) coated with a semi-reflecting film and the other half CD of the opposite face coated with a second film as nearly identical as possible with the first. In this system, each of the interfering beams is transmitted only once by plate S so that a compensator is not required; also, both beams are externally reflected at the beam-splitter so that the second cause of asymmetry is absent.

Control of path difference

A linear variation of path difference in time is not essential but the abscissae of the interferogram must be accurately known, the acceptable error in terms of wavelength being very small. This degree of accuracy cannot be met simply by displacing the moving mirror with a screw turning at a constant rate—the inevitable periodic variations in thread pitch would give rise to parasitic lines just as in the case of gratings. The situation in the present case is in fact worse than for gratings: in the Littrow mounting at an incident angle θ the relative intensity of a parasitic line due to a sinusoidal ruling error of amplitude ε is $(2\pi\varepsilon \sin \theta/\lambda)^2$ while in the Michelson interferometer the same error in the determination of path difference gives rise to a parasitic line of intensity $2\pi\varepsilon/\lambda$, which is considerably higher.

To determine the exact path difference continuously it is essential to make use of a reference fringe system of very high finesse generated by a beam passing through the interferometer in parallel with the radiation to be analysed. When the spectrum is calculated by a digital method the sinusoidal signal provided by the reference beam operates a punched-tape mechanism to mark those ordinates of the interferogram which correspond to equal increments of the path difference.

Since values of the transmitted flux are only read off point by point, it follows that the moving mirror need not be displaced continuously but can be given a stepping motion so that it moves quickly from one station to the next but is stationary while the measurement of transmitted flux is made. In this mode of operation the mirror displacement is again controlled by the reference signal.[29, 34, 64]

The stepping method offers several advantages. The equidistant spacing of the measurement stations can be achieved with an extremely high degree of accuracy, represented by a variation of not more than about 1 Å over the whole range of displacement.

Another advantage of the stepping method is that with

sources of variable intensity the duration of each measurement can be controlled by the flux (though Fourier spectroscopy of fluctuating sources is of course limited by the condition that variations in intensity of the source shall not change its spectrum). The output of an auxiliary detector (directly illuminated by the source) is integrated over a period of time inversely proportional to the mean source intensity at each setting of δ and the integrated signal operates a timing circuit.

7.41. Applications for Fourier spectroscopy

We recall that the essential feature of Fourier spectroscopy is the fact that radiations of all wavelengths in the spectrum to be measured are received simultaneously by the detector.

The simultaneous measurement of the intensities of N spectral elements provides an advantage that may be expressed in several ways:

> For a given resolving power and time of measurement the signal/noise ratio is multiplied by $\sqrt{N}$.
> For a given resolving power and signal/noise ratio, the time of measurement is divided by N.
> Other parameters being equal, there is a significant gain in resolving power.

The number N of spectral elements in a spectrum may be as high as, say, 10^6 which should, on the foregoing considerations, lead to a tremendous advantage for Fourier spectroscopy over other methods (the Fellgett advantage). In practice this advantage is only seen if the background noise in independent of the measured flux or, in other words, if the detector is the source of noise. This condition is met in the infra-red but not in the visible or ultra-violet regions, where noise originating in the photomultiplier is negligible compared with photon noise in the radiation.

It is therefore in the infra-red, where the Fellgett advantage holds, that Fourier spectroscopy finds its main justification.

Apart from this fundamental reason, practical considerations have also tended to promote the Fourier method as essentially a technique for the infra-red. It has indeed greatly facilitated the study of extremely weak infra-red spectra such as those encountered in astrophysics (for example, spectra of planets and the night sky) which can only otherwise be measured at very low resolving powers.

Although Fourier spectroscopy has this peculiar advantage in the infra-red, its potential in the visible and ultra-violet regions is also considerable, as P. Jacquinot has pointed out,[55] even though the Fellgett advantage is absent. Its strength here lies in the high angular aperture of the beam admitted and in the wide spectral range that can be covered in a single operation. This makes it possible to measure accurately the wavelengths of all the lines present by comparison with a single line of standard wavelength, a very valuable feature in atomic spectroscopy.

7.42. *Attainable performance and some results*

The refinements already described (cat's-eye reflectors and mirror spacing controlled by reference signal) have resulted in a reduction of the instrumental profile to a width of a few thousandths of a wavenumber, which in the near infra-red corresponds to a resolving power of 10^6. This demands a path difference variation of two metres, which in turn calls for reference fringes of very high finesse; these are therefore generated by laser radiation.[64]

P. and J. Connes[20, 22] have recently measured the spectra of some planets in three of the infra-red-transmitting spectral regions of the atmosphere, centred on 8000, 6000 and 4200 cm^{-1} (1·25 μm, 1·7 μm and 2·4 μm). Remarkably good results were obtained in spite of effects such as atmospheric turbulence which make astronomical observations difficult. The limit of resolution achieved—0·3 cm^{-1} for Jupiter, 0·08 cm^{-1} for Venus—is in both cases 100 times better than that of the best spectra previously recorded. A very large amount of information has, of course, been extracted from these spectra, including the very surprising discovery that the atmosphere of Venus contains hydrochloric and hydrofluoric acids.

Fourier spectroscopy is now a highly developed instrumental technique and its value has been fully established by the results obtained. Very high resolving powers can be attained but the capacity of present-day computers still limits the number of spectral elements that can be examined simultaneously. The full potential of Fourier spectroscopy will only be developed when advances in the computer programs or in the design of specialized high-speed computers make possible a significant reduction in the time taken for conversion of the interferogram into a spectrum.

7.43. Derivation of spectral profile from visibility of interference fringes

A two-beam interferometer yields an emergent flux expressed by the integral

$$\Phi(\delta) = k \int_{-\infty}^{+\infty} L_\sigma(\sigma)\{1 + \cos(2\pi\sigma\delta)\}\, d\sigma$$

Suppose the incident radiation to be quasi-monochromatic, extending over a very narrow band centred on a wavenumber σ_0.

Putting $\sigma' = \sigma - \sigma_0$ gives

$$\Phi(\delta) = k \int_{-\infty}^{+\infty} L_\sigma[1 + \cos\{2\pi(\sigma' + \sigma_0)\delta\}]\, d\sigma'$$

$$= k\bigg[L + \cos(2\pi\sigma_0\delta) \int_{-\infty}^{+\infty} L_\sigma \cos(2\pi\sigma'\delta)\, d\sigma'$$

$$- \sin(2\pi\sigma_0\delta) \int_{-\infty}^{+\infty} L_\sigma \sin(2\pi\sigma'\delta)\, d\sigma'\bigg]$$

L being the source radiance.

The above expression for $\Phi(\delta)$ can be written

$$\Phi(\delta) = k\{L + C\cos(2\pi\sigma_0\delta) - S\sin(2\pi\sigma_0\delta)\}$$

in which

$$C = \int_{-\infty}^{+\infty} L_\sigma \cos(2\pi\sigma'\delta)\, d\sigma'$$

$$S = \int_{-\infty}^{+\infty} L_\sigma \sin(2\pi\sigma'\delta)\, d\sigma'$$

whence

$$\Phi(\delta) = kL\left\{1 + \frac{\sqrt{(C^2+S^2)}}{L}\cos(2\pi\sigma_0\delta - \phi)\right\}$$

If we now limit the analysis to the case of a symmetrical spectral profile, $L_\sigma(\sigma')$ becomes an even function and the integral S is zero. The expression for the emergent flux is consequently reduced to

$$V(\delta) = \frac{C}{L} = \frac{\int_{-\infty}^{+\infty} L_\sigma \cos(2\pi\sigma'\delta)\, d\sigma'}{\int_{-\infty}^{+\infty} L_\sigma\, d\sigma'}$$

This shows that for a symmetrical profile the fringe visibility is proportional to the cosine Fourier transform of the function $L_\sigma(\sigma')$ describing that profile. ($V(\delta)$ may be regarded as the normalized Fourier transform of L_σ.) In this case a knowledge of $V(\delta)$ is sufficient to define the spectral profile; this is, however, not true for an asymmetric line.

J. Terrien, developing a method originated by Michelson, has shown that it is possible to derive precise information about the shape of a line directly from the $V(\delta)$ function without having to carry out a Fourier transformation.[72] This is the case, in particular, when a line is broadened by shock or by the Doppler–Fizeau effect.

Shock broadening gives a characteristic profile of the form

$$f_1(\sigma') = \frac{1}{1+(\sigma'^2/\Delta_1^2)} \quad \text{(Lorentz profile)}$$

Δ_1 being the half-width of the line.

Doppler–Fizeau broadening leads to a Gaussian profile

$$f_2(\sigma') = e^{-(\ln 2)\sigma'^2/\Delta_2^2}$$

where Δ_2 is again the half-width of the line.

In the first case the fringe visibility at a path difference δ is

$$V_1(\delta) = e^{-k_1\Delta_1\delta} \quad (\text{with } k_1 = +2\pi)$$

and in the second case

$$V_2(\delta) = e^{-k_2\Delta_2^2\delta^2} \quad (\text{with } k_2 = \pi^2/\ln 2)$$

The two causes considered are, in general, superimposed and

the resulting profile is the convolution of the two functions $f_1(\sigma')$ and $f_2(\sigma')$. The corresponding visibility is simply the product $V(\delta) = V_1 V_2$, so

$$V(\delta) = e^{-(k_1 \Delta_1 + k_2 \Delta_2^2 \delta)\delta}$$

and $(\ln V)/\delta = -k_1 \Delta_1 - k_2 \Delta_2^2 \delta$ is a linear function of the path difference, of which the slope gives the Doppler–Fizeau width Δ_2 and the ordinate at the origin the width Δ_1.

This method was chosen to select the most suitable radiation for definition of the unit of length.

7.44. *Fourier spectrography*

We have seen that in the case of Fourier spectrometry, the interferogram represents a Fourier transform of the spectrum in the temporal dimension, because the path difference varies as a function of time. It is precisely this translation into the time domain that makes it possible to record the interferogram with a flux detector. On the other hand the use of an image receptor would open up the possibility of a *spatial* display of the interferogram; this method has been studied for several years [75-78] and appears to offer certain advantages, notably the opportunity of utilizing an extended source of radiation.

Recording the interferogram

To display the Fourier transform of the spectrum as a spatial distribution of radiation it is clearly necessary to cause the source of radiation to form interference fringes, which may be, for example, Young's fringes. S. Lowenthal has described a system in which two identical images S_1 and S_2 of a broad source S (Fig. 7.29) are formed with a separation T by an image doubler D. A suitable device for this purpose is a Michelson interferometer in which each of the two mirrors is replaced by a pair of mirrors at right-angles, the vertices of these pairs being parallel. An objective O focuses the fringe system in its focal plane at H.

Any point M in the source S has two images at M_1 and M_2. The path difference $(MM_1 - MM_2)$ is independent of the position of M. Since the vector $\mathbf{M_1 M_2}$ is also constant for any position of M, the interference field at infinity is independent of the particular image pairs considered. Every point of the source therefore gives rise to the same system of fringes in the plane H, whatever the extent of the source. A monochromatic radiation of wavenumber σ thus forms a unique system of Young's fringes at H, of which the visibility is unity and is independent of the shape and size of the source. If Ω is the solid angle subtended by one of the source images at a point in the plane H,

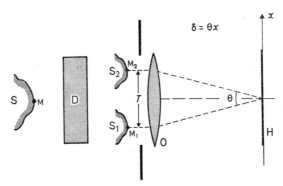

FIG. 7.29. Layout of system to produce hologram of a source S by means of Young's fringes.

the intensity at that point due to the integrated source radiations is

$$E(\delta) = \tfrac{1}{2}\Omega \int_0^\infty L_\sigma(\sigma)\{1+\cos(2\pi\sigma\delta)\}\,d\sigma$$

The variable term of this expression is, as in equation (7.19), proportional to the cosine Fourier transform of $L_\sigma(\sigma)$ and so, as before, represents the interferogram; in the present case, however, the path difference δ varies across the plane H so that a photographic plate mounted there records a *spatial* representation of the interferogram.

Reconstruction of the spectrum

There is no great difficulty in exposing and developing the plate under such conditions that the amplitude of radiation transmitted by the plate shall be proportional to $E(\delta)$ at every point. The spectrum can then be reconstructed by means of a simple Fraunhofer diffraction system or by using the holographic method of image retrieval. (The two methods are basically the same: the photographic record of the interferogram can be regarded as being a hologram.) So by illuminating the interferogram by a converging beam having a spherical wavefront, a distribution of radiation is obtained, of which the complex amplitude is proportional to the Fourier transform of $E(\delta)$. By reconstructing the spectral distribution with laser radiation, sufficient energy can be injected to allow the reconstructed spectrum to be scanned with a slit or pinhole much narrower than the width of the spread function associated with a given spectral element measured in the plane of the spectrum. It has been shown that this constitutes a filtering action which reduces the effect of emulsion noise.

Principal attributes of the method

The chief distinguishing feature of this technique lies in the way in which the interferogram is formed. The perfectly achromatic lateral image splitter makes it possible to produce a fringe system at infinity having unit visibility whatever the shape and size of the source. This advantage is fundamental since it ensures that resolving power and luminosity are independent of each other; the étendue of the instrument may, in theory, take any value. This must make the instrument particularly well suited to the analysis of very weak sources.

That this method is a form of Fourier spectroscopy is shown by the fact that each point in the plane H is irradiated by all the spectral elements constituting the sampled radiation while, inversely, the energy carried by each spectral element is spread over the whole plane. P. Bouchareine and P. Jacquinot[79] have shown that under these conditions the photographic emulsion is more efficiently utilized than in a spectrograph of the normal type and that the signal/noise ratio is enhanced. We have in addition noted that exploration of the spectrum with a very narrow aperture is possible and that this effectively reduces noise due to photographic grain.

The method, which is still undergoing development, seems

218   SPECTROSCOPY AND ITS INSTRUMENTATION

therefore to offer significant advantages for the spectroscopic analysis of sources of low radiance.

7.45. *Comparison of the various methods used in spectroscopy*

Fourier spectroscopy is essentially different from all the other methods discussed in that the spectrum is not recorded directly but has to be calculated by a time-consuming process. Nevertheless this complication is acceptable by virtue of the undoubted advantages offered by the method.

In these final paragraphs, we undertake a very general survey of the principles of operation of the various types of spectrometer. From this it will become apparent that the techniques employed for spectral analysis differ more in execution than in principle and that even Fourier spectroscopy has features in common with the other methods.[18]

7.46. *Autocorrelation function of luminous radiation*

The radiation analysed by a spectrometer is a fluctuating electric field and is a random function of time $E(t)$.† Let $C(\tau)$ be the autocorrelation function of $E(t)$.‡ It is shown in the theory of random functions that a function $L(\nu)$ representing the spectral power density can be associated with $E(t)$; this function is such that the power carried by radiation of frequencies within the band between ν and $\nu + d\nu$ is $L(\nu)\,d\nu$. This function $L(\nu)$ is recognizable as the radiation spectrum in the usual sense of this term. The determination of this function is the purpose of all instrumental spectroscopy.

Now, it can be shown that the spectrum $L(\nu)$ and the autocorrelation function of the radiation field are linked by a Fourier transformation:

---

† $E(t)$ is, more precisely, the analytic signal associated with the actual electric field.

‡ The autocorrelation function $C(\tau)$ of a function $y(t)$ is, by definition, the mean value (in the sense of mathematical expectation) of the product $y(t).y^*(t-\tau)$. In the cases of the functions considered here, it is identical with the mean calculated with reference to time, so that

$$C(\tau) = \overline{y(t).y^*(t-\tau)} = \lim_{T \to \infty} \frac{1}{2T} \int_{-T}^{T} y(t).y^*(t-\tau)\,dt$$

INTERFERENCE SPECTROSCOPY

$$L(\nu) = \int_{-\infty}^{+\infty} C(\tau) e^{-j2\pi\nu\tau} d\tau \qquad (7.30)$$

We shall now see how this function $C(\tau)$ occurs in spectral analysis.

7.47. *The case of Fourier spectroscopy*

The use of a two-beam interferometer provides the autocorrelation function $C(\tau)$ directly. This comes about because at the exit pupil of the instrument two vibrations of equal amplitude resulting from the splitting of the incident radiation are superimposed but one is retarded with respect to the other by a time $\tau = \delta/c$ (δ being the path difference). The resultant vibration is therefore:

$$F(t) = E(t) + E(t-\tau)$$

Its intensity (or mean power) is

$$I = \overline{F(t).F^*(t)} = \overline{E(t).E^*(t)} + \overline{E(t-\tau).E^*(t-\tau)}$$
$$+ \overline{E(t).E^*(t-\tau)} + \overline{E(t-\tau).E^*(t)}$$

The first two terms are equal to $C(0)$ and the other two to $C(\tau)$ and $C^*(\tau)$ respectively. We thus have

$$I = 2C(0) + 2C_R(\tau) \qquad (7.31)$$

in which $C_R(\tau)$ is the real part of the autocorrelation function $C(\tau)$.

In Fourier spectroscopy the interferogram (the variable component of the intensity I) therefore represents the function $C_R(\tau)$.

The autocorrelation function $C(\tau)$, being the Fourier transform of a real function, will have a real part C_R which is an even function and an imaginary part C_I which is an odd function. Equation (7.30) can therefore be written

$$L(\nu) = \int_{-\infty}^{+\infty} C_R(\tau)\cos(2\pi\nu\tau) d\tau + \int_{-\infty}^{+\infty} C_I(\tau)\sin(2\pi\nu\tau) d\tau \qquad (7.32)$$

or
$$L(\nu) = L_a(\nu) + L_b(\nu),$$

$L_a(\nu)$ being even and $L_b(\nu)$ odd.

But $L(\nu)$ is zero for $\nu < 0$. In this case we have, for positive values of ν,
$$L_a(\nu) = L_b(\nu) = \tfrac{1}{2}L(\nu)$$

Knowing the function $L_a(\nu)$ is therefore equivalent to knowing $L(\nu)$ and, since

$$L_a(\nu) = \int_{-\infty}^{+\infty} C_R(\tau)\cos(2\pi\nu\tau)\, d\tau$$

the fact that the cosine Fourier transform of the interferogram gives the spectrum is confirmed.

The autocorrelation function of the electromagnetic field fluctuation is therefore a valid starting point for an analysis of the Fourier transform method of spectroscopy.

7.48. *The case of other methods*

In contradistinction to Fourier spectroscopy, which is based on two-beam interferometry, all other methods rely on some form of multiple-beam interferometry (Fabry–Perot étalon, grating, or prism†). The incident wavetrain gives rise to N wavetrains by separation in space (grating) or in amplitude (Fabry–Perot), having amplitudes $a_0, a_1, \ldots, a_n$ of which the retardations relative to the first are, respectively, $0, \theta, 2\theta, \ldots$. The emergent wavetrain resulting from their superposition is expressed by

$$F(t) = \sum_{n=0}^{N} a_n E(t - n\theta) \tag{7.33}$$

From now on, we shall take the summation to infinity even though N may in real cases be limited as it is, for instance, with gratings; in such cases, we can put $a_n = 0$ for $n > N$.

---

† It has already been pointed out that the principle of a prism spectrometer is based on an interference effect.

INTERFERENCE SPECTROSCOPY

The quantity measured at a given setting of the instrument, that is, for a particular value θ of the optical lag τ between two successive wavetrains, is the intensity I of the emergent radiation.

The variations of I as a function of θ are taken as representing the required spectrum $L(\nu)$. It remains, then, to establish the relationship between the two functions $I(\theta)$ and $L(\nu)$.

From equation (7.33)

$$I = \overline{F(t).F^*(t)} = \sum_{p,q} a_p a_q \overline{E(t-p\theta).E^*(t-q\theta)}$$

(the a coefficients are real).

Collecting the terms for which $|p-q| = n$; we have

$$I = (a_0^2 + a_1^2 + \cdots)C(0) \\ + \sum_{n=1}^{\infty}(a_0 a_n + a_1 a_{n+1} + \cdots)[C(n\theta) + C(-n\theta)]$$

Writing the real part of $C(\tau)$ as $C_R(\tau)$, this reduces to

$$I = \tfrac{1}{2}\mathscr{A}_0 C(0) + \sum_{n=1}^{\infty} \mathscr{A}_n C_R(n\theta) \qquad (7.34)$$

with

$$\mathscr{A}_n = 2(a_0 a_n + a_1 a_{n+1} + \cdots) = 2\sum_{p=0}^{\infty} a_p a_{n+p}$$

The intensity I of the emergent radiation is thus seen to be the sum of a series of terms which are the values assumed by the correlation function $C_R(\tau)$ of the radiation at equally spaced values $\tau = n\theta$, each term being first multiplied by its own factor $\mathscr{A}_n$.

Instead of the retardation θ, we can substitute the corresponding value $h = c\theta$ of the path difference δ. The autocorrelation function $C_R(\tau)$ is then replaced by $\mathscr{C}_R(\delta)$. We also introduce a function $\mathscr{A}(\delta)$ which assumes values $\mathscr{A}_n$ for $\delta = nh$. This now gives

$$I = \tfrac{1}{2}\mathscr{A}(0)\mathscr{C}(0) + \sum_{n=1}^{\infty} \mathscr{A}(nh)\mathscr{C}_R(nh) \qquad (7.35)$$

for the emergent radiation intensity.

Returning now to equation (7.29) used to derive the spectrum from the Fourier spectrometer interferogram and putting $\sigma = k/h$, h being an integer, we get

$$F'(k/h) = h[\tfrac{1}{2}H(0)\mathscr{I}(0) + H(h)\mathscr{I}(h) + H(2h)\mathscr{I}(2h) + \cdots]$$
(7.36)

since cosines are now all equal to unity.† This expression is analogous to equation (7.35), $H(\delta)$ and $\mathscr{I}(\delta)$ having the same roles as their respective counterparts $\mathscr{A}(\delta)$ and $\mathscr{C}_R(\delta)$; this emphasizes the correspondence between Fourier spectroscopy and those methods that utilize multiple-beam interferometry.

In the former case the recorded interferogram approximately represents the correlation function of the radiation and the Fourier transform is calculated from sampled values.

In the latter case the interferometer carries out the sampling automatically. When it is adjusted so that the path difference between two wavetrains is h, the resultant radiant flux is proportional to the value assumed by the Fourier transform of $\mathscr{A}(\delta)\mathscr{C}(\delta)$ when $\sigma = k/h$. (This value of σ is, of course, the setting wavenumber.) The two classes of method are therefore completely analogous.

In the two cases, the sampling of the correlation function at non-zero intervals h results in the situation that there is no unique spectrum: there are as many superimposed spectra as there are possible values of k.‡

It must, however, be particularly noted that the correlation function $\mathscr{C}_R(\delta)$ is only known over a finite interval δ_M since it is always multiplied by a limiting function which is zero or nearly so when $\delta > \delta_M$. Thus in Fourier spectrometry the interferogram is only recorded for $\delta < \delta_M$. In other methods the factor $\mathscr{A}(\delta)$ expresses the fact that the number of lines in a grating is limited or that the multiple reflections within a Fabry–Perot étalon decrease in amplitude. The Fourier transform of $\mathscr{A}(\delta)$ can also be shown to be the instrumental profile $A(\sigma)$.

---

† The integer k introduced here is the order of interference, since h, the path difference between the successive wavetrains, is $k\sigma$.

‡ The exceptional case is that of prism instruments, for which $h = 0$.

7.49. Conclusion

The purpose of spectroscopic instrumentation may be summed up as being the determination of the Fourier transform of the autocorrelation function $\mathscr{C}_R(\delta)$ of the radiation under investigation, since this transform is identical with the spectrum $L(\nu)$.

The function $\mathscr{C}_R(\delta)$ is synthesized from the output of a two-beam or a multiple-beam interferometer. In the first case, which is that of a Fourier spectrometer, the autocorrelation function is recorded directly and its Fourier transform calculated afterwards. In all the other methods the value of the Fourier transform of $\mathscr{C}_R(\delta)$ at each wavenumber setting is obtained directly.

In practice, the autocorrelation function is never known exactly. This factor of approximation is reflected in the existence of an instrumental profile which always has a finite width and in a resolving power which is always limited.

BIBLIOGRAPHY

CHAPTER 2

1. P. BOUCHAREINE et P. JACQUINOT, Spectrographie par grille, *J. Phys.*, 1967, **28**, Suppl. C2, 183–193.
2. C. CANDLER, *Modern interferometers*, Adam Hilger (London, 1951), p. 436.
3. J. DUFAY, *Introduction à l'astrophysique: les étoiles*, Armand Colin (Paris, 1961), p. 218.
   P. B. FELLGETT, Design of astronomical telescopes, in J. HOME DICKSON (ed.), *Optical instruments and techniques*, Oriel Press (Newcastle-upon-Tyne, 1970), pp. 483–6.
4. P. JACQUINOT et CH. DUFOUR, Conditions optiques d'emploi des cellules photoélectriques dans les spectrographes et les interféromètres, *J. Rech. Cent. nat. Rech. Scient.*, 1948, **2**, 91–103.
5. A. MARÉCHAL et M. FRANÇON, *Diffraction, structure des images, influence de la cohérence de la lumière*, published by la Revue d'Optique (Paris, 1960), p. 204.
   M. BORN and E. WOLF, *Principles of optics*, Pergamon Press (London, third revised edition 1965), chapter 10.

CHAPTER 3

1. A. ARNULF, Spectrographes à grande ouverture applicables à l'ultraviolet et à l'infrarouge, *Ann. Astrophys.*, 1943, **6**, 21–35.
2. J. CABANNES, Les spectrographes modernes et leurs applications à la Physique moléculaire, *Rev. Scient.*, 1943, **81**, 419–432.
3. J. COJAN, Le spectrographe du Palais de la Découverte, *Rev. Opt. Théor. Instrum.*, 1945, **24**, 35.
4. J. COJAN, Un nouveau spectrographe très ouvert, *Ann. Astrophys.*, 1947, **10**, 33–38.
5. A. COUDERC, Construction et essai d'un grand spectrographe à liquide. *J. Phys. Radium*, 1937, Série VII, **8**, 99 S.

Translator's note. For further reading on prism spectographs, see R. A. SAWYER, *Experimental spectroscopy*, Dover Publications (New York, 1963), pp. 50–126.

CHAPTER 4

1. H. D. BABCOCK and H. W. BABCOCK, The ruling of diffraction gratings at the Mount Wilson Observatory, *J. Opt. Soc. Am.*, 1951, **41**, 776.
2. H. W. BABCOCK, Control of a ruling engine by a modulated interferometer, *Appl. Opt.*, 1962, **1**, 415–429.
3. P. BOUSQUET, Etude théorique de la diffraction de la lumière par les réseaux. *Proceedings of the conference on photographic and spectroscopic optics* (Tokyo, 1964). *Jap. J. Appl. Phys.*, 1965, **4**, Suppl. I, 549–554.

4. A. Camus (Melle), M. Françon, E. Ingelstam et A. Maréchal, Etude des erreurs de tracé des réseaux par contraste de phase. *Rev. Opt. Théor. Instrum.*, 1951, **30**, 121–137.
5. R. Deleuil, Etude expérimentale de la diffraction des ondes électromagnétiques par les réseaux, *C. R. hebd. Séanc. Acad. Sci.*, 1966, **262**, 1676–1679.
6. G. D. Dew and L. A. Sayce, On the production of diffraction gratings: (I) The copying of plane gratings, *Proc. Roy. Soc.*, 1951, A **207**, 278.
7. R. G. N. Hall and L. A. Sayce, On the production of diffraction gratings: (II) The generation of helical rulings and the preparation of plane gratings therefrom, *Proc. Roy. Soc.*, 1959, **215** A, 536.
8. G. R. Harrison, The production of diffraction gratings: (I) Development of the ruling art, *J. Opt. Soc. Am.*, 1949, **39**, 413–426.
9. G. R. Harrison, The production of diffraction gratings: (II) The design of echelle gratings and spectrographs. *J. Opt. Soc. Am.*, 1949, **39**, 522–528.
10. G. R. Harrison and G. W. Stroke, Interferometric control of grating ruling with continuous carriage advance, *J. Opt. Soc. Am.*, 1955, **45**, 112–121.
11. G. R. Harrison, N. Sturgis, S. C. Baker and G. W. Stroke, Ruling of large diffraction gratings with interferometric control, *J. Opt. Soc. Am.*, 1957, **47**, 15–22.
12. G. R. Harrison, N. Sturgis, S. P. Davis and Yahiko Yamada, Interferometrically controlled ruling of ten-inch diffraction gratings, *J. Opt. Soc. Am.*, 1959, **49**, 205–211.
13. G. R. Harrison and G. W. Stroke, Attainment of high resolution with diffraction gratings and echelles, *J. Opt. Soc. Am.*, 1960, **50**, 1153–1158.
14. E. Ingelstam and E. Djurle, The study of diffraction grating characteristics by simplified phase contrast methods, *J. Opt. Soc. Am.*, 1953, **43**, 572.
15. R. F. Jarrel, 'Gratings, production of' in the *Encyclopedia of spectroscopy*, Reinhold Publishing Corp. (New York, 1960), p. 173.
16. A. Keith Pierce, Performance of an eight-inch Babcock grating in a large vacuum spectrograph. *J. Opt. Soc. Am.*, 1957, **47**, 6.
17. R. P. Madden and J. Strong, Appendix P in J. Strong, *Concepts of classical optics*, Freeman and Co. (San Francisco, 1958), p. 605.
18. R. P. Madden, Preparation and measurement of reflecting coatings for the vacuum ultra-violet in G. Hass, *Physics of thin films*, Academic Press (New York, 1963), **1**, pp. 123–186.
19. A. Maréchal and G. W. Stroke, Sur l'origine des effets de polarisation et de diffraction dans les réseaux optiques, *C. R. hebd. Séanc. Acad. Sci.*, 1959, **249**, 2042–2044.
20. Sir Thomas Merton, On the reproduction and ruling of diffraction gratings, *Proc. Roy. Soc.*, 1950, **201** A, 187.
21. R. Petit, Contribution à l'étude de la diffraction d'une onde plane par un réseau métallique, *Rev. Opt. Théor. Instrum.*, 1963, **42**, 263–281.
22. R. Petit, Diffraction d'une onde par un réseau métallique, *Rev. Opt. Théor. Instrum.*, 1966, **45**, 249–276, 353–370.
23. D. H. Rank, J. N. Shearer and J. M, Bennett, Quantitative method for measuring the resolution of a large grating, *J. Opt. Soc. Am.*, 1965, **45**, 762.
24. D. H. Rank, A. H. Guenther, C. R. Burnett and T. A. Wiggins, Examples of high resolution obtainable by double passing a high quality grating, *J. Opt. Soc. Am.*, 1957, **47**, 631.

25. D. RICHARDSON, Modern diffraction gratings, *Proceedings of the conference on photographic and spectroscopic optics* (Tokyo, 1964), *Jap. J. Appl. Phys.*, 1965, **4**, Suppl. I, 562–563.
26. G. W. STROKE, Interferometric measurement of wave-front aberrations in gratings and echelles, *J. Opt. Soc. Am.*, 1955, **45**, 30–35.
    F. SIMEON, in C. CANDLER, *Modern interferometers*, Adam Hilger (London, 1954), pp. 151–152.
27. G. W. STROKE, Etudes théorique et expérimentale de deux aspects de la diffraction par les réseaux optiques: l'évolution des défauts dans les figures de diffraction et l'origine électromagnétique de la répartition entre les ordres, *Rev. Opt. Théor. Instrum.*, 1960, **39**, 291.
28. G. W. STROKE, Attainment of high-resolution gratings by ruling under interferometric control, *J. Opt. Soc. Am.*, 1961, **51**, 1321–1339.

Translator's note. For further reading on aberrations, see W. T. WELFORD, Aberration theory of gratings and grating mountings, in E. WOLF (ed.), *Progress in optics*, North Holland (Amsterdam, 1965), Vol. IV, pp. 243–280.

29. G. W. STROKE and H. H. STROKE, Tandem use of gratings and echelles to increase resolution, luminosity and compactness of spectrometers and spectrographs, *J. Opt. Soc. Am.*, 1963, **53**, 333–338.
30. G. W. STROKE, Attainment of high efficiencies in blazed optical gratings by avoiding polarisation in the diffracted light, *Physics Letters*, 1963, **5**, 45–48.
31. G. W. STROKE, Ruling, testing and use of optical gratings for high resolution spectroscopy, in *Progress in optics*, North Holland (Amsterdam, 1963), Vol. II, pp. 1–72.
32. G. W. STROKE, Diffraction gratings, in *Handbuch der Physik* (Springer-Verlag), Vol. 29, pp. 426–754.
33. J. STRONG, The Johns Hopkins University and diffraction gratings, *J. Opt. Soc. Am.*, 1960, **50**, 1148–1162.
34. LORD RAYLEIGH, *Proc. Roy. Soc.*, 1872, **20**, 414–417.
35. A. A. MICHELSON, *Studies in optics*, University Press (Chicago, 1917).
36. A. LABEYRIE and J. FLAMAND, Spectrographic performance of holographically-made diffraction gratings, *Optics communications*, 1969, **1**, 5–8.
37. G. PIEUCHARD, J. FLAMAND, J. CORDELLE et A. LABEYRIE, Réseaux de diffraction holographiques, Presented at a Colloquium in Marseilles (September, 1969), organized by the Centre National d'Etudes Spatiales.
38. J. CORDELLE, J. FLAMAND, G. PIEUCHARD and A. LABEYRIE, Aberration-corrected concave gratings made holographically, in J. HOME DICKSON (ed.), *Optical instruments and techniques*, Oriel Press (Newcastle-upon-Tyne, 1970), pp. 117–124.

CHAPTER 5

1. N. ASTOIN, Spectrographe à réseau concave en incidence rasante dans le vide pour l'ultraviolet lointain, *J. Phys. Radium*, 1951, **12**, 695–696.
2. A. BAYLE, J. ESPIARD, C. BRETON, M. CAPET et L. HERMAN, Le spectrographe échelle R.E.O.S.C. type H.A., *Rev. Opt. Théor. Instrum.*, 1962, **41**, 585–593.
3. A. BARANNE, Thesis, Marseille, 1965.
4. H. EBERT, *Wied. Ann.*, 1889, **38**, 489.

5. Ch. FEHRENBACH, Le nouveau spectrographe installé au foyer coudé du télescope de 1,93 m de l'Observatoire de Haute Provence, Pub. OHP, 1959, Vol. 5, No. 1. For further details, see G. COURTÈS, Ch. FEHRENBACH, E. HUGHES et J. ROMAND, Quelques réalisations instrumentales en France, *Appl. Opts*, 1966, **5**, 1351.
6. G. R. HARRISON, The production of diffraction gratings: (II) The design of echelle gratings and spectrographs, *J. Opt. Soc. Am.*, 1949, **39**, 522–528.
7. G. R. HARRISON, J. E. ARCHER and J. CAMUS, A fixed-focus broad range echelle spectrograph of high speed and resolving power, *J. Opt. Soc. Am.*, 1952, **42**, 706–712.
8. G. R. HARRISON and G. W. STROKE, Attainment of high resolution with diffraction gratings and echelles, *J. Opt. Soc. Am.*, 1960, **50**, 1153–1158.
9. R. F. JARRELL, Stigmatic plane grating spectrograph with order sorter. *J. Opt. Soc. Am.*, 1955, **45**, 259–269.
10. D. H. RANK, A. H. GUENTHER, C. R. BURNETT and T. A. WIGGINS, Examples of high resolution obtainable by double passing a high quality grating, *J. Opt. Soc. Am.*, 1957, **47**, 631–635.
11. Mme S. ROBIN et S. ROBIN, Spectrographe à vide et reseau concave en incidence normale pour l'ultraviolet lointain, *J. Phys. Radium*, 1956, **17**, 976.
12. Mme S. ROBIN et S. ROBIN, Spectrographe à réseau pour l'ultraviolet lointain, *Rev. Opt. Théor. Instrum.*, 1958, **37**, 161–170.
13. J. ROMAND et B. VODAR, Un spectrographe dans le vide pour l'analyse spectrochimique d'émission dans l'ultraviolet lointain, *Rev. Opt. Théor. Instrum.*, 1958, **37**, 329–335.
14. C. J. SILVERNAIL, High speed wide-range fluorite spectrograph, *J. Opt. Soc. Am.*, 1957, **47**, 23–26.
15. G. W. STROKE, Ruling, testing and use of optical gratings for high resolution spectroscopy, *Progress in optics*, North Holland (Amsterdam, 1963), Vol. II, pp. 3–72.
16. G. W. STROKE and H. H. STROKE, Tandem use of gratings and echelles to increase resolution, luminosity and compactness of spectrometers and spectrographs, *J. Opt. Soc. Am.*, 1963, **53**, 333–338.

Translator's note. For further reading on far ultra-violet spectroscopy see R. TOUSEY, The extreme ultra-violet—past and future, *Appl. Opt.*, 1962, **1**, 679–694.

CHAPTER 6

1. R. F. BAKER, *J. Opt. Soc. Am.*, 1938, **28**, 55.
2. S. S. BALLARD, Infra-red optical materials, old and new. *Proceedings of the conference on photographic and spectroscopic optics*. Tokyo, 1964. *Jap. J. Appl. Phys.*, 1965, **4**, Suppl. I, 23–29.
3. W. G. FASTIE, A small plane grating monochromator, *J. Opt. Soc. Am.*, 1952, **42**, 641–647.
4. W. G. FASTIE, Image forming properties of the Ebert monochromator, *J. Opt. Soc. Am.*, 1952, **42**, 647–651.
5. W. G. FASTIE, H. M. CROSSWHITE and P. GLOERSEN, Vacuum Ebert grating spectrometer, *J. Opt. Soc. Am.*, 1958, **48**, 106–111.
6. A. GIRARD, Etude d'un spectromètre à modulation sélective, *Publications de l'Office National d'Etudes et de Recherches Aérospatiales*, n° 117 (1967).

7. A. HADNI, Contribution à l'étude théorique et expérimentale de l'infrarouge lointain, *Ann. Physique*, 13ᵉ Série, 1956, **1**, 234–290.
8. A. HADNI, J. CLAUDEL, E. DECAMPS, X. GERBAUX et P. STRIMER, Spectres d'absorption de monocristaux dans l'infrarouge lointain (50–1 600 μ), à la température de l'hélium liquide, *C. R. hebd. Séanc. Acad. Sci.*, 1962, **255**, 1595–1597.
9. P. JACQUINOT et Ch. DUFOUR, Conditions optiques d'emploi des cellules photoélectriques dans les spectrographes et les interféromètres, *J. Rech. Cent. Nat. Rech. Scient.*, 1948, **2**, 91–103.
10. P. JACQUINOT, Luminosités comparées des spectromètres à prismes et à réseaux, *Rev. Opt. Théor. Instrum.*, 1954, **33**, 653–658.
11. P. JACQUINOT, The luminosity of spectrometers with prisms, gratings or Fabry–Perot etalons, *J. Opt. Soc. Am.*, 1954, **44**, 761–765.
12. J. A. MUIR and R. J. CASHMAN, A new infra-red-transmitting chalcogenide glass, *J. Opt. Soc. Am.*, 1967, **57**, 1–3.
13. T. NAMIOKA, Theory of the concave grating, I, *J. Opt. Soc. Am.*, 1959, **49**, 446–460. Theory of the concave grating, II: Application of the theory of the off-plane Eagle mounting in a vacuum spectrograph, *J. Opt. Soc. Am.*, 1959, **49**, 460–465.
14. T. NAMIOKA, Theory of the concave grating III: Seya–Namioka monochromator, *J. Opt. Soc. Am.*, 1959, **49**, 951–961.
15. T. NAMIOKA, Design of high-resolution monochromator for the vacuum ultraviolet. An application of off-plane Eagle mounting, *J. Opt. Soc. Am.*, 1959, **49**, 961–965.
16. M. POUEY et J. ROMAND, Monochromateur pour la région 400 à 3000 Å, *Rev. Opt. Théor. Instrum.*, 1965, **44**, 445–458.
17. S. ROBIN, Méthode de focalisation d'un monochromateur pour l'ultraviolet lointain muni de fentes fixes et distantes, *J. Phys. Radium*, 1953, **14**, 551–552.
18. J. ROMAND et B. VODAR, Un monochromateur dans le vide pour l'ultraviolet lointain, *Rev. Opt. Théor. Instrum.*, 1960, **39**, 167–174.
19. J. ROMAND et B. VODAR, Un monochromateur à réseau concave en incidence tangentielle pour l'ultraviolet lointain, *Optica Acta*, 1962, **9**, 371–381.
20. Y. SAKAYANAGI, Transmission gratings as an infrared filter, *Proceedings of the conference on photographic and spectroscopic optics*, Tokyo, 1964, *Jap. J. appl. Phys.*, 1965, **4**, Suppl. I, 397–400.
21. T. SAKURAI and S. TAKAHASHI, Far infra-red spectrometer with a variable depth grating. *Proceeding of the conference on photographic and spectroscopic optics*, Tokyo, 1964. *Jap. J. appl. Phys.*, 1965, **4**, Suppl. I, 358–363.
22. M. SALLE et B. VODAR, Projet de réalisation d'un monochromateur à réseau concave en incidence oblique, pour l'ultraviolet lointain, *C. R. hebd. Séanc. Acad. Sci.*, 1950, **230**, 380–382.
23. M. SEYA, *Sc. Light*, 1952, **2**, 8.
24. J. STRONG, Interferometric modulator, *J. Opt. Soc. Am.*, 1954, **44**, 352.
25. R. TOUSEY, F. S. JOHNSON, J. RICHARDSON and N. TORAN, A monochromator for the vacuum ultra-violet, *J. Opt. Soc. Am.*, 1951, **41**, 696–698.

CHAPTER 7

General references

Colloque international sur les progrès récents en spectroscopie interférentielle, Bellevue, 1957. *J. Phys. Radium*, 1958, **19**, 185–436.
P. JACQUINOT, New developments in interference spectroscopy. Reports on *Progress in Physics*, 1960, **23**, 267–312.
Colloque sur les méthodes nouvelles de spectroscopie instrumentale, Orsay, 1966. *J. Phys.*, 1967, **28**, Suppl. C_2, 1–344.
A. GIRARD and P. JACQUINOT, Principles of instrumental methods in spectroscopy, in A. C. S. VAN HEEL (ed.), *Advanced optical techniques*, North Holland (Amsterdam, 1967).
G. VANASSE and H. SAKAI, Fourier spectroscopy, in E. WOLF (ed.) *Progress in optics*, North Holland (Amsterdam, 1967), Vol. IV, pp. 261–330.

1. J. BLAISE, Description du spectromètre Fabry-Perot enregistreur de Bellvue, *J. Phys. Radium*, 1958, **19**, 335–337.
2. J. BLAISE, Recherches sur le déplacement isotopique dans les spectres atomiques des éléments lourds, *Annls. Phys.*, 1958, **3**, 1019–1076.
3. D. J. BRADLEY, Parallel movement for high finesse interferometric scanning, *J. Scient. Instrum.*, 1962, **39**, 41–45.
4. D. J. BRADLEY, B. BATES, C. O. JUULMAN and S. MAJUMDAR, The Fabry–Perot interferometer in the middle and vacuum ultra-violet. *Proceedings of the conference on photographic and spectroscopic optics*, Tokyo, 1964. *Jap. J. Appl. Phys.*, 1967, **4**, Suppl. I, 467–472.
5. D. J. BRADLEY, B. BATES, C. O. JUULMAN and T. KOHNO, Recent developments in the application of the Fabry–Perot interferometer to space research. *Colloque sur les méthodes nouvelles de spectroscopie instrumentale*. *J. Phys.* 1967, **28**, Suppl. C2, 280–286.
6. R. CHABBAL, Recherche des meilleures conditions d'utilisation d'un spectromètre photoélectrique Fabry–Perot, *J. Rech. Cent. Nat. Rech. Scient.*, 1953–54, **5**, 138–186.
7. R. CHABBAL, Recherches expérimentales sur la généralisation de l'emploi du spectromètre Fabry–Perot aux divers domaines de la spectroscopie, *Revue Opt. Théor. Instrum.*, 1958, **37**, 49–103, 336–370, 501.
8. R. CHABBAL, Le spectromètre Fabry–Perot intégral, *J. Phys. Radium*, 1958, **19**, 246–255.
9. R. CHABBAL et M. SOULET, Dispositif permettant le déplacement mécanique d'une lame de Fabry-Perot, *J. Phys. Radium*, 1958, **19**, 274–277.
10. R. CHABBAL, Finesse limite d'un Fabry–Perot formé de lames imparfaites, *J. Phys. Radium*, 1958, **19**, 295–300.
11. R. CHABBAL et P. JACQUINOT, Description d'un spectromètre interférentiel Fabry–Perot, *Revue Opt. Théor. Instrum.*, 1961, **40**, 157–170.
12. R. CHABBAL et R. PELLETIER, Principe et réalisation d'un spectromètre Fabry–Perot multicanal: le S.I.M.A.C. *Proceedings of the conference on photographic and spectroscopic optics*, Tokyo, 1964. *Jap. J. Appl. Phys.*, 1965, **4**, Suppl. I, 445–447.
13. R. CHABBAL, Ph. BIED-CHARRETON et R. PELLETIER, Le S.I.M.A.C.; utilisation avec une plaque photographique ou une caméra électronique. *Colloque sur les méthodes nouvelles de spectroscopie instrumentale*. *J. Phys.*, 1967, **28**, Suppl. C2, 209–214.

14. H. CHANTREL, Un double étalon à balayage par pression, *J. Phys. Radium*, 1958, **19**, 366–370.
15. H. CHANTREL, Recherches sur la structure hyperfine des spectres atomiques, et sur la structure fine de l'hélium au moyen de spectromètres Perot–Fabry enregistreurs à un ou deux étalons, *Annls. Phys.*, 1959, **4**, n°s 7/8.
16. H. CHANTREL, Spectromètres interférentiels de haute résolution à un ou deux étalons de Fabry–Perot, *J. Rech. Cent. Nat. Rech. Scient.*, 1959, **46**, 17–33.
17. H. CHANTREL, J. L. COJAN et P. GIACOMO, Couches réfléchissantes multidiélectriques pour l'ultraviolet. Mesures interférentielles sur la structure de la raie de résonance du mercure. *Colloque international sur l'Optique des couches minces solides*, Marseille, 1963. *J. Phys.*, 1964, **25**, 280–284.
18. J. CONNES, Recherches sur la spectroscopie par transformation de Fourier, *Revue Opt. Théor. Instrum.*, 1961, **40**, 45–79, 116–140, 171–190 et 231–265.
19. J. CONNES et V. NOZAL, Le filtrage mathématique dans la spectroscopie par transformation de Fourier, *J. Phys. Radium*, 1961, **22**, 359.
20. J. CONNES, P. CONNES et J. P. MAILLARD, Spectroscopie astronomique par transformation de Fourier. *Colloque sur les méthodes nouvelles de spectroscopie instrumentale. J. Phys.*, 1967, **28**, Suppl. C2, 120–135.
21. J. CONNES et P. CONNES, Méthodes de calcul digital. *Colloque sur les méthodes de spectroscopie instrumentale. J. Phys.*, 1967, **28**, Suppl. C2, 57.
22. J. CONNES and P. CONNES, Near-infra-red planetary spectra by Fourier spectroscopy, *J. Opt. Soc. Am.*, 1966, **56**, 896–910.
23. P. CONNES, Augmentation du produit luminosité × résolution des interféromètres par l'emploi d'une différence de marche indépendante de l'incidence, *Revue Opt. Théor. Instrum.*, 1956, **35**, 37.
24. P. CONNES, Spectromètre interférentiel à sélection par l'amplitude de modulation, *J. Phys. Radium*, 1958, **19**, 215–222.
25. P. CONNES, L'étalon de Perot–Fabry sphérique, *J. Phys. Radium*, 1958, **19**, 262–269.
26. P. CONNES, Principe et réalisation d'un nouveau type de spectromètre interférentiel, *Revue Opt. Théor. Instrum.*, 1959, **38**, 157, 416; 1960, **39**, 402.
27. P. CONNES, Décomposition des raies spectrales par modulation en haute fréquence, *J. Phys. Radium*, 1962, **23**, 173–183.
28. P. CONNES, DUONG HONG TUAN et J. PINARD, Détection d'un faible effet Doppler par l'emploi de franges de superposition, *J. Phys. Radium*, 1962, **23**, 208.
29. P. CONNES et J. PINARD, Procédés d'enregistrement pas à pas des interférogrammes. *Colloque sur les méthodes nouvelles de spectroscopie instrumentale.* Orsay, 1966.
30. J. COOPER and J. R. GREIG, Rapid scanning of spectral line profiles using an oscillating Fabry–Perot interferometer, *J. Scient. Instrum.*, 1963, **40**, 433–437.
31. G. COURTÈS, Méthodes d'observation et étude de l'hydrogène interstellaire en émission, *Ann. Astrophys.*, 1960, **23**, 115–217.
32. G. COURTÈS, Etude de l'émission interstellaire à l'aide de l'étalon de Perot–Fabry, *J. Phys. Radium*, 1958, **19**, 342–345.
33. G. COURTÈS et Y. GEORGELIN, Montage Perot–Fabry à multilentilles *Colloque sur les méthodes nouvelles de spectroscopie instrumentale. J. Phys.*, 1967, **28**, Suppl. C2, 218–220.

34. M. Cuisenier et J. Pinard, Spectromètre de Fourier à 'œil de chat' et à balayage rapide. *Colloque sur les méthodes nouvelles de spectroscopie instrumentale. J. Phys.*, 1967, **28**, Suppl. C2, 97–104.
35. Ch. Dufour et R. Picca, Sur l'interféromètre Fabry–Perot: Importance des imperfections des surfaces, *Revue Opt. Théor. Instrum.*, 1945, **24**, 19.
36. Ch. Dufour, Recherches sur la luminosité, le contraste et la résolution de systèmes interférentiels à ondes multiples: Utilisation des couches minces complexes, *Annls. Phys.*, 1951, **6**, 5.
37. R. Dupeyrat, Etude de procédés électriques de balayage pour des interféromètres enregistreurs, *J. Phys. Radium*, 1958, **19**, 290–292.
38. R. F. Edgar, B. Lawrenson and J. Ring, An optical analogue Fourier transformer. *Colloque sur les méthodes nouvelles de spectroscopie instrumentale. J. Phys.*, 1967, **28**, Suppl. C2, 73–78.
39. P. Fellgett, A propos de la théorie du spectromètre interférentiel multiplex, *J. Phys. Radium*, 1958, **19**, 187–191.
40. S. Fujita and H. Yoshinaga, A new computing method for interference spectroscopy. *Proceedings of the conference on photographic and spectroscopic optics*, Tokyo, 1964. *Jap. J. Appl. Phys.*, 1965, **4**, Suppl. I, 429–432.
41. P. Giacomo, Les couches réfléchissantes multidiélectriques appliquées à l'interféromètre de Perot–Fabry. Etude théorique et expérimentale des couches réelles. *Revue Opt. Théor. Instrum.*, 1956, **35**, 317–354, 442–467.
42. P. Giacomo, Propriétés chromatiques des couches réfléchissantes multidiélectriques. *J. Phys. Radium*, 1958, **19**, 307–311.
43. P. Giacomo, Applications des couches minces en optique. *Colloque internationale sur l'optique des couches minces solides*, Marseille, 1963. *J. Phys.*, 1964, **25**, 238–244.
44. G. Graner, Propriétés et réalisation d'un spectromètre S.I.S.A.M. dans l'infrarouge proche. *J. Phys.*, 1965, **26**, 222 A–228 A.
45. G. Graner, High resolution absorption spectroscopy in the infra-red with Fabry–Perot and S.I.S.A.M. spectrometers, *Appl. Opt.*, 1965, **4**, 1620–1623.
46. H. P. Gush and H. L. Buijs, High resolution Fourier transform spectroscopy. *Colloque sur les méthodes nouvelles de spectroscopie instrumentale. J. Phys.*, 1967, **28**, Suppl. C2, 105–108.
47. G. Henderson, H. T. Betz and P. N. Slater, Applications of the PRISM Fabry–Perot in solar physics and airglow. *Colloque sur les méthodes nouvelles de spectroscopie instrumentale. J. Phys.*, 1967, **28**, Suppl. C2, 287–288.
48. J. G. Hirschberg and P. Platz, A multichannel Fabry–Perot interferometer, *Appl. Opt.*, 1965, **4**, 1375.
49. J. G. Hirschberg, Recent developments in the application of the multichannel Fabry–Perot to plasma spectroscopy. *Colloque sur les méthodes nouvelles de spectroscopie instrumentale. J. Phys.*, 1967, **28**, Suppl. C2, 226–229.
50. D. A. Jackson, The spherical Fabry–Perot interferometer as an instrument of high resolving power for use with external or internal atomic beam, *Proc. Roy. Soc.*, 1961, A **263**, 289–308.
51. P. Jacquinot et Ch. Dufour, Conditions optiques d'emploi des cellules photoélectriques dans les spectrographes et les interféromètres, *J. Rech. Cent. Nat. Rech. Scient.*, **2** (1948), 91–103.

BIBLIOGRAPHY 233

52. P. JACQUINOT, Quelques perspectives d'avenir en spectrographie instrumentale. XVIIe Congrès du Groupement pour l'Avancement des Spectroscopiques, Paris, 1954.
53. P. JACQUINOT, The luminosity of spectrometers with prisms, gratings or Fabry–Perot etalons, J. Opt. Soc. Am., 1954, **44**, 761–765.
54. P. JACQUINOT, Caractères communs aux nouvelles méthodes de spectroscopie interférentielle; facteur de mérite, J. Phys. Radium, 1958, **19**, 223–229.
     P. B. FELLGETT, On the theory of infra-red sensitivities and its application to the investigation of stellar radiation in the near infra-red. PhD. Thesis, University of Cambridge, 1951.
55. P. JACQUINOT, Progrès récents en spectroscopie interférentielle. Proceedings of the conference on photographic and spectroscopic optics, Tokyo, 1964. Jap. J. Appl. Phys., 1965, **4**, Suppl. I, 401–411.
56. F. A. JENKINS, Extension du domaine spectral de pouvoir réflecteur élevé des couches multiples diélectriques, J. Phys. Radium, 1958, **19**, 301–306.
57. G. I. KATCHEN, J. KATZENSTEIN and L. LOVISETTO, A multi-channel photoelectric spectrometer employing a Fabry–Perot etalon and axicon. Colloque sur les méthodes de spectroscopie instrumentale. J. Phys., 1967, **28**, Suppl. C$_2$, 230–237.
58. J. KATZENSTEIN, The axicon-scanned Fabry–Perot spectrometer, Appl. Opt., 1965, **4**, 263.
59. J. KUHL, A. STEUDEL and H. WALTHER, A digital recording double Perot–Fabry Spectrometer. Colloque sur les méthodes nouvelles de spectroscopie instrumentale. J. Phys., 1967, **28**, Suppl. C2, 308–312.
60. J. P. LAUDE, Spectromètre à deux Fabry-Perot asservis. Colloque sur les méthodes nouvelles de spectroscopie instrumentale. J. Phys., 1967, **28**, Suppl. C2, 322–325.
61. R. LENNUIER, Réalisation de miroirs interférentiels pour le domaine ultraviolet ($\lambda = 2500$ Å), J. Phys. Radium, 1958, **19**, 319–320.
62. J. E. MACK, D. P. MCNUTT, F. L. ROESLER and R. CHABBAL, The P.E.P.S.I.O.S. purely interferometric high resolution scanning spectrometer, Appl. Opt., 1963, **2**, 873–885.
63. M. MORILLON (Mme), Mise au point et performances d'un spectromètre interférentiel à sélection par l'amplitude de modulation (S.I.S.A.M.) fonctionnant en absorption dans la région de 5 μ. Colloque sur les méthodes nouvelles de spectroscopie instrumentale. J. Phys., 1967, **28**, Suppl. C2, 181–182.
64. J. PINARD, Spectromètre de Fourier à très haute résolution. Colloque sur les méthodes nouvelles de spectroscopie instrumentale. J. Phys., 1967, **28**, Suppl. C2, 136–143.
65. P. PLATZ et J. G. HIRSCHBERG, Etude de la température Doppler dans une décharge toroïdale à l'aide d'un interféromètre Perot–Fabry multicanal. C. r. hebd. Séanc. Acad. Sci., 1965, **261**, 1207–1210.
66. J. L. PRITCHARD, A. BULLARD, H. SAKAI and G. A. VANASSE, Idealab Fourier transform analog computer. Colloque sur les méthodes nouvelles de spectroscopie instrumentale. J. Phys., 1967, **28**, Suppl. C2, 67–72.
67. J. V. RAMSAY, A rapid-scanning Fabry–Perot interferometer with automatic parallelism control, Appl. Opt., 1962, **1**, 411–413.
68. J. V. RAMSAY, Automatic control of Fabry–Perot interferometers. Colloque sur les méthodes nouvelles de spectroscopie instrumentale. J. Phys., 1967, **28**, Suppl. C2, 321.

69. F. L. ROESLER and J. E. MACK, The P.E.P.S.I.O.S. spectrometer. *Colloque sur les méthodes de spectroscopie instrumentale. J. Phys.*, 1967, **28**, Suppl. C2, 313–320.
70. P. N. SLATER, H. T. BETZ and G. HENDERSON, A new design of a scanning Fabry–Perot interferometer, *Proceedings of the conference on photographic and spectroscopic optics*, Tokyo, 1964. *Jap. J. Appl. Phys.*, 1965, **4**, Suppl. I, 440–444.
71. A. STEUDEL, Préparation et propriétés de couches réfléchissantes pour le Fabry–Perot dans l'ultraviolet, *J. Phys. Radium*, 1958, **19**, 312–318.
72. J. TERRIEN, J. HAMON et T. MASUI, Profil spectral et causes d'élargissement de quelques radiations hautement monochromatiques du mercure 198, *C. r. hebd. Séanc. Acad. Sci.*, 1957, **245**, 926–929.
73. J. VERGÈS, Fonction d'appareil et performances d'un S.I.S.A.M. à haute résolution. *Colloque sur les méthodes nouvelles de spectroscopie instrumentale. J. Phys.*, 1967, **28**, Suppl. C2, 176–180.
74. H. YOSHINAGA, Recent techniques in far infra-red spectroscopy, *Proceedings of the conference on photographic and spectroscopic optics*, Tokyo, 1964. *Jap. J. Appl. Phys.*, 1965, **4**, Suppl. I, 420–428.
    H. YOSHINAGA, see S. FUJITA and YOSHINAGA, reference 40.
75. G. W. STROKE and A. T. FUNKHOUSER, Fourier-transform spectroscopy using holographic imaging without computing and with stationary interferometers, *Physics Letters*, 1965, **16**, 272–274.
76. J. Ch. VIÉNOT et G. PERRIN, Transmission des hologrammes au moyen d'un chaîne de télévision, *C.R. Acad. Sc. Paris*, 1968, **267 B**, 1137–1140.
77. R. PRAT, *Jap. J. Appl. Phys.*, Suppl. I, 1965, **4**, 448.
78. S. LOWENTHAL, C. FROEHLY et J. SERRES, Spectrographie de Fourier à haute luminosité et faible bruit par application des techniques holographiques, *C.R. Acad. Sc. Paris*, 1969, **268 B**, 1481–1484.
79. P. BOUCHAREINE et P. JACQUINOT, *J. Physique*, 1967, **28**, Suppl. C2, 183–193.
80. J. L. RAYCES, Formation of Axicon images, *J. Opt. Soc. Am.*, 1958, **48**, 576–78.

INDEX

Aberrations in spectrographs, 34, 82
—in spectrometers, 113–14, 118–19, 124
Absorption spectra, measurement of, 135–7
Airy function for étalon, 152–3, 171
Aluminium film for gratings, 65
—, reflectivity of, u.v., 73–4
Aperture of spectrograph, 24, 34
—, relative, of collimator, 34–5
—, —, of focusing lens, 22–3, 33
—, resolving, 22–4
Apodization of Fourier spectrometer, 202
— of SISAM, 194–5
Arsenic trisulphide, 109
Astigmatism in spectrometers, 119, 124
Astronomy, Fabry–Perot étalon in, 188–90
—, Fourier spectrometer in, 212
—, spectrograph for, 82–4
Autocorrelation function of radiation, 218–23
Axicon lens, 181

Bandwidth of monochromator, 140–1, 143
Blaze angle, 52
— effect, 52

Caesium bromide, 110
— iodide, 110
Calcium aluminate, 109
— fluoride, 31, 110
Classification of dispersive instruments 5–6
Coating, high reflection, metallic, 172
—, —, multilayer, 172–4
Coherent illumination, 12
Collimator, spectrograph, 34–5
Coma in spectrograph mounting, 82
— in spectrometer mounting, 113–14, 118–19
Connes spherical étalon, 181–4
Cornu prism, 31
Criterion of resolution, 10–11, 146
Crystals for infra-red prisms, 110–11

Detector noise, in Fourier spectrometer, 211
— —, in grille spectrometer, 134–5

Diffraction, effect on instrumental profile, 102
—, — on resolving power, 8–10
— field, form of, 40–8
Doppler effect in nebulae, 188–90

Échelette profile, 52, 59–61
Échelle grating, 50, 85
— spectrographs, 84–9
Échelon, Michelson, 50
Emulsion, limit of resolution, 21
—, photographic, effect of, 21–7
Equation, grating, 47–8
Étalon, Fabry–Perot, 148–90
—, —, Airy function of, 152–3
—, —, application of, 150–1
—, —, as monochromator, 184
—, —, defect finesse of, 155–8, 160
—, —, diaphragm aperture, effect of, 158
—, —, high-reflection coatings for, 171–4
—, —, in astronomy, 188–90
—, —, instrumental profile of, 152–60
—, —, monitor for parallelism, 169–70
—, —, optical principles, 149–50
—, —, practical features of, 171–4
—, —, reflective finesse of, 154–5, 161–2
—, —, reflectivity of plates, 152–5, 171–4
—, —, spectrometer, see Spectrometer, Fabry–Perot étalon
—, —, spherical, 181–4
—, —, —, luminosity of, 182
—, —, transmission factor of, 150, 162, 171–3
Étendue, definition of, 4 (footnote)
— of spectrometer, Fabry–Perot, 163
— —, Fourier, 203–5
— —, grille, 133–4
— —, slit, 104
— of spherical étalon, 182

Fabry–Perot étalon, see Étalon, Fabry–Perot
— — spectrometer, see Spectrometer, Fabry–Perot
— interferometer, 148–90

Fellgett advantage, 196 (footnote), 211
Finesse, coefficient of, 146–8
—, defect, 155–8
—, reflective, 154–5
Filters, infra-red, 121–2
Fluorite, 31, 110
Fourier spectrography, see Spectrography, Fourier
— spectrometer, see Spectrometer, Fourier
— spectrometry, principle of, 197–201, 219–20
Free spectral range, 50–1, 121, 147
Fringes, coefficient of finesse of, 146
—, visibility of, 213–15

Germanium, 111, 174
Ghost lines, 55–8, 68–9
— —, intensity of, 57–8
Glass for prisms, 30
— —, infra-red, 109
Grain, photographic, 21–7
Grating defects, 53–63
—, diffraction by, 40–78
—, efficiency of, 59–60, 77
— equation, 47–8
—, Harrison's échelle, 50
— materials, 64–5
—, Michelson échelon, 50
— pitch, 51
—, resolving power, 48–50
—, Siegbahn, 76
Gratings as order separators, 121
—, blazed, 52
—, concave, 122–5
—, —, stigmatic, 78
— échelette, 52, 59–61
— échelle, 85
— effect of interferometric control, 68
— for far ultra-violet, 73–6
—, holographic, 76–8
—, —, efficiency of, 77
— methods of inspection, 61–3
— methods of production, 63–78
— — —, interferometric, 65–8
— replica, 71–3
— variable-depth, 119–20
Grille spectrograph, 26–7
— spectrometer, 125–35
— —, alternation system, 130–1
— —, étendue advantage, 133–4
— —, instrumental profile, 128–33

Grille spectrometer, oscillation system, 131–3
— —, slit function, 125–6

Hydrogen H_α line, 188–90
HYPEAC spectrometer, 176–8

Incoherent illumination, 12
Infra-red, gratings for, 73
—, order separators for, 121–2
—, prisms for, 109–111
—, spectrometers for, 109, 211–12
—, surface coatings for, 174
Instrumental profile, Fabry–Perot spectrometer, 152–60
— —, — —, effect of diaphragm aperture, 158
— —, — —, effect of surface form, 155–8
— —, — —, effect of surface reflectivity, 154–5
— —, — —, inclusive, 158–61
— —, Fourier spectrometer, 201–2, 203–5, 207
— —, grille spectrometer, 128–33
— —, monochromator, 139–40
— —, SISAM, 193–4
— —, slit spectrometer, 96–102
Interferogram, 198, 205–8, 215–16
Interferometer as dispersing instrument, 145–95
—, coefficient of finesse of, 147–8
—, Fabry–Perot, 148–90
—, —, optical principles, 149–50
—, —, theoretical resolving power, 146–7
—, free spectral range of, 147–8
—, SISAM, 190–5
IRTRAN materials, 111

Jacquinot advantage, 165, 205

Kayser, 3
KRS-5, 110

Length, primary standard of, 2
Lens, collimator, 34–5
—, focusing, 33–4
Line, spectral, profile of, 213–15
Lithium fluoride, 31
$\mathscr{LR}$ invariant, 106, 159, 164–5
Luminosity, 5
—, effect of slit width on, 19–20

INDEX

Luminosity of monochromator, 140–1
— of SISAM, 194
— of spectrograph, 19, 22–7, 103
— —, stellar, 26
— of spectrometer, 99, 102–9, 162–4
— of spherical étalon, 182

Merton nut, 72
— replica gratings, 71–3
Mirror, off-axis paraboloid, 116–17
Monochromator, SISAM, 194
—, spherical étalon, 184
Monochromators, slit, 138–43
—, —, additive mounting, 141–2, 143
—, —, bandwidth of, 140–1, 143
—, —, double, 141–3
—, —, instrumental profile of, 139–140
—, —, luminosity of, 140–1
—, —, subtractive mounting, 142–3
Mounting, Arnulf, 34
—, constant-deviation, 111
—, classical prism, 35
—, Czerny–Turner, 113, 117
—, Eagle, 91–3
—, —, modified, 93
—, —, off-plane, 93
—, Ebert, 82
—, Ebert–Fastie, 118–19
—, étalon spectrograph, classical, 184–5
—, grazing-incidence, 94
—, Littrow, 35–6, 114, 119, 131
—, monochromator, additive, 141–3
—, —, subtractive, 142–3
—, multiple-grating, 87–9
—, multiple-pass, 87
—, Paschen–Runge, 91
—, Pellin–Broca, 113
—, plane grating, normal, 80–1
—, Pouey–Romand–Martin, 124–5
—, Rowland, 91
—, Seya–Namioka, 123–4
—, Wadsworth, 112
—, Z, 82, 113–14
Multiplexing, principle of, 196

Nebulae, Doppler effect in, 188–90
Noise, see Detector noise and signal/noise ratio
Nut, Merton, 72
—, Rowland, 63

Optical acceptance, 5
Order separators, 84, 85, 120–2, 165–8
— —, far infra-red, 121–2

Partial coherence, 13
Plate, photographic, grain of, 21
Prism angle, choice of, 31–2
—, Cornu, 31
—, dimensions, choice of, 32
—, liquid, 33
— materials, 29–31
— —, infra-red, 109–11
—, resolving power of, 29
— train, 33
Profile, échelette, 52
—, ruling, 51–2, 59–61, 62–3

Quartz for prisms, 30–1
— —, infra-red, 109

Rayleigh criterion, 11
Reflectivity, boundary, 32
—, ultra-violet, of metals, 73–6
Refractive index of IRTRAN materials 111
Replica gratings, 71–3
Resolution, criterion of, 10–11, 146
—, definition of, 4
Resolvable spectral element, 4
Resolvance, see Resolving power
Resolving aperture, 22
Resolving power, definition, 4
— —, effect of slit width on, 19–20
— —, grain-limited, 22–3
— — of dispersing element, 8–12
— — —, intrinsic, 11–12
— — of grating, 48–50, 70
— — of interferometer, 146–7
— — of prism, 29
— — of SISAM, 194
— — of spectrograph, 8–20, 103
— — of spectrometer, 98, 102–9, 159
— — —, Fourier, 201–3
Rowland circle, 90, 122, 123
— nut, 63
Ruling engines, 63–8
—, interferometric control of, 65–8
— errors, 55–63
— profile, 51–2, 59–61
— tolerances, 58

Sampling interval, 207–8
— theorem, 206

Satellite lines, 59, 69
Scanning, methods of, for Fabry–Perot étalon, 168–71
Scattered light from gratings, 59
Siegbahn grating, 76
Signal/noise ratio, 5, 134, 196, 211
Silica, fused, for prisms, 30–1
—, —, —, infra-red, 109
Silicon, 111, 174
SIMAC spectrograph, 186–8
SISAM spectrometer, *see* Spectrometer, SISAM
Slit function of grille spectrometer, 125–6
— illumination, effect of coherence, 12–13
— image, distribution of intensity in, 13–19
— width, monochromator, 143
— —, spectrograph, effect on luminosity, 19–20
— —, —, effect on resolving power, 19–20
— —, spectrometer, effect on instrumental profile, 99–102
— — —, effect on luminosity, 102–6
— —, —, effect on resolving power, 102–3
Sodium chloride, 110
Source function, 96
—, radiance of, 3
Spectra, absorption, measurement of, 135–7
—, superposition of, 50–1, 120–2
Spectral element, resolvable, 4
Spectral line width, 213–15,
Spectrograph, basic layout of, 7
—, definition of, 6
—, échelle, 84–9
—, Fabry–Perot étalon, 184–5
—, — —, classical mounting, 184–5
—, high-aperture, 37–8
—, low-luminosity, 37
—, medium-power, 38
— mounting, Arnulf, 34
— —, classical, 35, 80–1
— —, Eagle, 91–3
— —, Ebert, 82
— —, grazing-incidence, 94
— —, Littrow, 35–6
— —, multiple-grating, 87–9
— —, multiple-pass, 87

Spectrograph mounting, Paschen–Runge, 91
— —, Rowland, 91
— —, Z, 82
—, SIMAC, 186–8
—, stellar, luminosity of, 26
Spectrographs, comparison with spectrometers, 102–3
—, concave grating, 90–4
—, grating, mountings for, 80–2, 87–9, 91–4
—, plane grating, 79–89
—, luminosity of, 19, 22–7, 103
—, multiple-camera, 82–4
—, prism, mountings for, 35–6
—, —, performance of, 39
—, resolving power of, 8–27
—, — —, effective, 19, 25
Spectrography, Fourier, 215–18
—, grille, 26–7
Spectrometer, definition of, 6
—, Fabry–Perot étalon, 151–84
—, — —, advantages of, 165
—, — —, choice of parameters for, 160–2
—, — —, comparison with grating spectrometer, 164–5
—, — —, étendue of, 163
—, — —, fast-scanning, 179–80
—, — —, for hyperfine line structure, 175–8
—, — —, HYPEAC, 176–8
—, — —, instrumental profile of, 152–60
—, — —, Jacquionot advantage of, 165
—, — —, $\mathscr{LR}$ invariant for, 164–5
—, — —, luminosity of, 159, 162–4
—, — —, multichannel, 180–1
—, — —, resolving power of, 159
—, — —, separation of orders, 165–8
—, — —, single-band, 178–9
—, — —, spectrum scanning methods for, 168–71
—, Fourier, 196–218
—, —, apodization of, 202
—, —, applications of, 211–12
—, —, control of path difference, 210–11
—, —, computation of spectrum, 205–8
—, —, design of, 208–11

INDEX 239

Spectrograph, Fourier, étendue of, 203–5
—, —, instrumental profile of, 201–2, 203–5, 207
—, —, performance of, 212–13
—, —, principle of, 197–201
—, —, resolving power of, 201–3
—, —, symmetrical form of, 209
—, grille, 125–35
—, —, alternation system, 130–1
—, —, detectors for, 134–5
—, —, étendue advantage of, 133–4
—, —, instrumental profile of, 128–33
—, —, mounting for, 131
—, —, oscillation system, 131–3
—, —, slit function, 126
— mountings, 111–16, 117–19, 122–5
— —, constant-deviation, 111
— —, Czerny–Turner, 113, 117
—, —, Ebert–Fastie, 118–19
— —, Littrow, 114, 119, 131
— —, Pellin–Broca, 113
— —, Pouey–Romand–Martin, 124–5
— —, Seya–Namioka, 123–4
— —, Wadsworth, 112
— —, Z, 113–14
—, SISAM, 190–5
—, —, apodization of, 194–5
—, —, applications of, 194, 195
—, —, construction of, 195
—, —, instrumental profile of, 193–4
—, —, luminosity of, 194
—, —, resolving power of, 194
Spectrometers, comparison with spectrographs, 102–3
—, grating, 117–25
—, — and prism compared, 107–9
—, —, astigmatism in, 119, 124
—, —, concave, 122–5
—, —, —, astigmatism in, 122–5
—, —, mountings for, 117–19, 122–5

Spectrometers, prism, 109–17
—, — and grating compared, 107–9
—, —, constant-deviation, 111
—, —, mountings for, 111–16
—, slit, instrumental profile of, 96–102
—, —, luminosity of, 99, 102–9
—, —, properties of, 95–109
—, —, resolving power of, 98–9, 102–9
—, —, — —, definition, 98–9
—, —, recorded function, 97
Spectrophotometers, double-beam, 136–7
Spectroscopy, comparison of methods, 218–23
—, Fourier, 196–218, 219–20
Spectrum, computation from interferogram, 205–8
—, high-speed scanning of, 179–80
—, order of, 49
—, reconstruction of, from spatial interferogram, 216–17
Speculum metal, 64
Standard of length, wavelength, 2

Terminology for spectroscopic instruments, 5–6
Thallium bromoiodide, 110

Ultra-violet, dielectric multilayers for, 174
—, grating for far, 73–6, 94
—, spectrometers for far, 122–5

Wavelength, definition of, 1
—, international, 2
— standard, primary, 2
—, units of, 1
Wavenumber, definition of, 3
—, setting, definition of, 95